Statistics for Biology and Health

Series Editors
K. Dietz, M. Gail, K. Krickeberg, J. Samet, A. Tsiatis

T0214387

Springer

London
Berlin
Heidelberg
New York
Hong Kong
Milan
Paris
Tokyo

Statistics for Biology and Health

Borchers/Buckland/Zucchini: Estimating Animal Abundance: Closed Populations.

Everitt/Rabe-Hesketh: Analyzing Medical Data Using S-PLUS.

Ewens/Grant: Statistical Methods in Bioinformatics: An Introduction.

Hougaard: Analysis of Multivariate Survival Data.

Klein/Moeschberger: Survival Analysis: Techniques for Censored and Truncated Data.

Kleinbaum: Logistic Regression: A Self-Learning Text, 2nd ed.

Kleinbaum: Survival Analysis: A Self-Learning Text.

Lange: Mathematical and Statistical Methods for Genetic Analysis, 2nd ed.

Manton/Singer/Suzman: Forecasting the Health of Elderly Populations.

Salsburg: The Use of Restricted Significance Tests in Clinical Trials.

Sorensen/Gianola: Likelihood, Bayesian, and MCMC Methods in Quantitative Genetics.

Therneau/Grambsch: Modeling Survival Data: Extending the Cox Model.

Zhang/Singer: Recursive Partitioning in the Health Sciences.

D.L. Borchers, S.T. Buckland and
W. Zucchini

Estimating Animal Abundance

Closed Populations

With 91 Figures

 Springer

Series Editors

K. Dietz
Institut für Medizinische Biometrie
Universität Tübingen
West Banhofstrasse 55
D-72070 Tübingen
GERMANY

M. Gail
National Cancer Insttitue
Rockville, MD 20892
USA

K. Krickeberg
Le Chatelet
F-63270 Manglieu
FRANCE

A. Tsiatis
Department of Statistics
North Carolina State University
Raleigh, NC 27695
USA

J. Samet
School of Public Health
Department of Epidemiology
Johns Hopkins University
615 Wolfe Street
Baltimore, MD 21205-2103
USA

British Library Cataloguing in Publication Data
Borchers, D. L.
   Estimating animal abundance : closed populations. – (Statistics for biology and health)
   1. Animal populations - Statistical methods 2. Estimation theory
   I. Title II. Buckland, S. T. (Stephen T.) III. Zucchini, W.
   591.7'88'0727

Library of Congress Cataloging-in-Publication Data
Estimating animal abundance : closed populations / D.L. Borchers, S.T. Buckland, and W. Zucchini.
   p.   cm. -- (Statistics for biology and health)

   1. Animal populations - Statistical methods. I. Borchers, D.L., 1958-
II. Buckland, S.T. (Stephen T.) III. Zucchini, W. IV. Series.
QL752.E76 2002
591.7'88—dc21   2002021159

Apart from any fair dealing for the purposes of research or private study, or criticism or review, as permitted under the Copyright, Designs and Patents Act 1988, this publication may only be reproduced, stored or transmitted, in any form or by any means, with the prior permission in writing of the publishers, or in the case of reprographic reproduction in accordance with the terms of licences issued by the Copyright Licensing Agency. Enquiries concerning reproduction outside those terms should be sent to the publishers.

ISBN 978-1-84996-885-0
Springer-Verlag is a part of Springer Science+Business Media
springeronline.com

© Springer-Verlag London Limited 2010
Printed in Great Britain
2nd printing 2004

The use of registered names, trademarks, etc. in this publication does not imply, even in the absence of a specific statement, that such names are exempt from the relevant laws and regulations and therefore free for general use.

The publisher makes no representation, express or implied, with regard to the accuracy of the information contained in this book and cannot accept any legal responsibility or liability for any errors or omissions that may be made.

12/3830-54321 Printed on acid-free paper

Preface

We hope this book will make the bewildering variety of methods for estimating the abundance of animal populations more accessible to the uninitiated and more coherent to the cogniscenti. We have tried to emphasize the fundamental similarity of many methods and to draw out the common threads that underlie them. With the exception of Chapter 13, we restrict ourselves to closed populations (those that do not change in composition over the period(s) being considered). Open population methods are in many ways simply extensions of closed population methods, and we have tried to provide the reader with a foundation on which understanding of both closed and open population methods can develop.

We would like to thank Miguel Bernal for providing the St Andrews example dataset used frequently in the book; Miguel Bernal and Jeff Laake for commenting on drafts of the book; Jeff Laake for providing Figure 10.1; NRC Research Press for allowing us to use Figures 10.2, 10.3, 10.4, 10.5, 10.6 and 10.7; the International Whaling Commission for allowing us to use Figure 12.1; Sharon Hedley for providing Figures 12.1 and 12.2.

D.L.B. is eternally indebted to Carol, Alice and Aidan for their support through writing the book, and for the many evenings and weekends that it has taken from them.

D.L.B. and S.T.B. acknowledge the financial support of the University of St Andrews.

W.Z. gratefully acknowledges the financial support of the University of Gottingen, Martin Erdelmeier's enormous contributions to the development of WiSP, and the boundless moral support provided by Leon and Laura.

David Borchers, Stephen Buckland and Walter Zucchini

Contents

II Simple Methods

III   Advanced Methods

IV Overview

Part I

Introduction

1
Introduction

Figure 1.1 shows the result of survey conducted to estimate the number of plants in a 50m by 100m field. The dots represent detected plants. An eighth of the survey region (the shaded rectangles) was searched so if we were sure every plant in the shaded rectangles was detected, it would be reasonable to assume that about an eighth of the plants in the field were detected. This would lead us to estimate that there were 200 plants present.

In practice you never know what proportion of the population you have detected. Often some are missed even in the searched region, and so estimating the probability of detection is a key part of estimating abundance. The number of ways of doing this can be quite daunting, but almost all

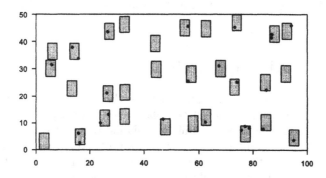

Figure 1.1. Plants (dots) detected by searching the shaded plots, which comprise an eighth of the survey region.

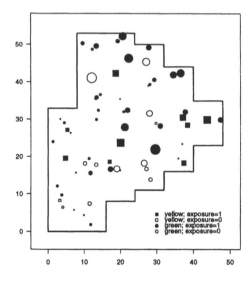

Figure 1.2. Plants of a particular species detected by an observer in a field. Symbol size is proportional to plant size. Plants of this species have either yellow or green flowers and plants are classified as "exposure=1" if they are clearly exposed above surrounding vegetation, and as "exposure=0" otherwise.

of them can be viewed as variants or combinations of only three basic approaches.

We illustrate the approaches by example below, using the data in Figure 1.2. It shows 64 plants of a particular species that were detected by an observer walking through an irregularly shaped field. What proportion of the population did she detect?

The observer feels that she saw about half the plants. If we accept this, we would estimate that there were 128 plants in the field. But we don't have good reason to make this assumption (her feeling may be quite wrong). It would be better to get an estimate that was based more on data and less on assumption.

1.1 Estimation approach 1

Realizing that if we sent enough observers through the field they must between them eventually see all the plants, we send a pair of observers through the field. They see 100 different plants. If they have now seen all the plants, then sending a team of still more observers in will lead to no more than 100 detections. To check we send a five-observer team through. They see 147. Is this the whole population in the field? A plot of total number of plants detected against number of observers (Figure 1.3) can help answer this. It suggests that the number of detections approaches

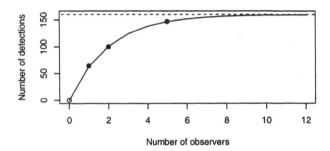

Figure 1.3. Number of plants detected as a function of number of observers used. The solid line is the fitted curve, the dashed line is the resulting estimate of abundance, the solid circles are the observed numbers.

some maximum as the number of observers increases – as we'd expect – the maximum being the population size.

We could keep sending more and more observers until increasing the number of observers leads to no increase in the number of plants seen; the number seen at this point would be our estimate of abundance. Alternatively we could estimate abundance by fitting a smooth curve to these points and finding the point at which the curve is flat (i.e. the point at which we predict that adding more observers would add no more detections). This curve is shown in Figure 1.3. The circle at (0,0) is not an observation, but clearly with no observers there will be no detections.

Figure 1.3 gives an estimate of the proportion of the plants detected (height of solid line divided by height of dashed line), as a function of number of observers. Using the figure, we estimate that the first observer detected about 40% of the population and that there are about 160 plants in the population. We have used an observable feature of the survey (number of observers) to estimate the proportion of plants detected. We could only do this because we surveyed with different numbers of observers and we believe that with enough observers all plants are detected.

In a similar way, we could use a feature of the plants rather than a feature of the survey to estimate the proportion detected. Distance sampling methods do this; we cover them in Chapter 7.

1.2   Estimation approach 2

The second approach is based on the insight that the rate at which the number of new detections falls off after removal of some plants tells us something about the proportion of the population that was removed. Suppose each observer removed all the plants that they detected. If a large proportion was removed by the first observer, and the second observer applied the same search effort, they would detect far fewer plants than the

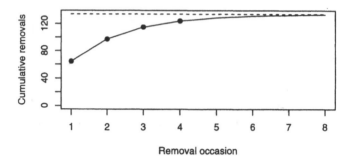

Figure 1.4. Cumulative total of plants detected and removed over the first four survey occasions. The solid line is the fitted curve, the dashed line is the resulting estimate of abundance, the solid circles are the observed numbers.

first. If the first observer removed a very small fraction of the plants present, the second observer would detect about the same number of plants as the first observer.

Figure 1.4 shows the cumulative removals of the first four observers. The numbers of plants detected and removed by these observers are: 64, 33, 18 and 9, giving cumulative removals of 64, 97, 115 and 124. Each observer detected substantially fewer plants than the previous observer, suggesting that a substantial fraction of the remaining population was removed on each occasion.

With enough removal occasions, all the plants will be removed and at this point we will know what the population size was. But this might take a huge amount of effort and we might not want to remove the whole population! So instead of removing all the plants, we can fit a line to the observed series of cumulative removals and predict when no plants will be left. This is the point at which additional removal occasions yield no more removals, i.e. the point at which the fitted line is horizontal. The fitted line is shown in Figure 1.4 and from it we estimate that there were initially about 134 plants in the population. The estimated proportion of the population that was detected by observer 1 is the height of the solid line on removal occasion 1 divided by the height of the dashed horizontal line, i.e. about 48%.

There is a variety of estimation methods constructed using this approach, often using a version of method 1 above as well. Collectively they are known as "removal methods". When they also treat the proportion detected as a function of number of observers (as above, for example) or any measure of survey effort, they are called "catch-effort" methods. When they use information on removals of different plant or animals types (sex for example), they are called "change-in-ratio" methods. These methods are usually used for animals, not plants.

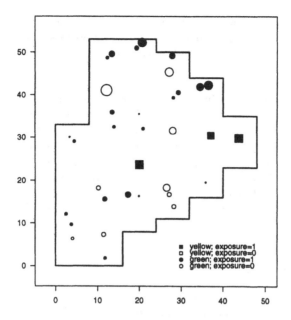

Figure 1.5. Plants of a particular species in a field detected by both the first and the second observer who surveyed the field. Symbol size is proportional to plant size.

1.3   Estimation approach 3

The third approach is possibly the simplest conceptually, and yet it has spawned the most complicated variety of sub-methods. Figure 1.5 shows the plants detected by the first observer that were also detected by the second observer. There are 34 of them. That is, the second observer detected 34 64ths (about 53%) of the plants known to be present because the first observer saw them. From this we estimate that the second observer detected about 53% of the population in all. The number of "recaptures" (34 in this case) contains information about the proportion of plants observer 2 detected. This idea is the basis of mark-recapture methods. Using these data, and the fact that observer 2 saw 67 plants in total, we estimate that there are 126 plants in the field.

1.4   Heterogeneity

In the discussion above, we ignored the fact that some plants may be more detectable or more catchable than others. Look at Figures 1.2 and 1.5 again. Some plants are large, some small; some have yellow flowers, some green; some are exposed above surrounding vegetation, some are not. Each of these features might influence detectability. By comparing the sorts of plants that were seen by both observers with the sorts seen by one observer, you might get some idea of how plant properties affect detection probability. For example, do the plots suggest that plants with green flowers are more detectable?

We refer to differences in detectability or catchability due to features of the animals or plants themselves as "heterogeneity". Ignoring it can result in large bias: if all observers saw only the very detectable animals, the abundance estimate would be the abundance of very detectable animals, not of all animals. Dealing with heterogeneity appropriately can be surprisingly difficult and in most cases requires fairly sophisticated statistical modelling. We deal first with the simpler methods, to a large extent leaving heterogeneity to the "Advanced Methods" part of the book.

1.5 Summary

For the most part, there are only three basic approaches underlying the many methods of estimating the number of animals in a closed population. They

(1) use the change in the number of detections as some observable feature of the survey or the objects changes, or they

(2) use the change in the number of detections as the population is reduced by removing objects, or they

(3) use the proportion of objects that are recaptured after being captured and marked.

Line transect and point transect methods are examples of approach 1. Removal, catch-effort and change-in-ratio methods are examples of methods that use approach 2. Mark-recapture methods use approach 3. Some methods used more than one approach.

You will have noticed that the three methods we used to estimate plant abundance in the field produced different estimates. For any given population, some methods will give more reliable estimates than others. In the chapters that follow, we consider the circumstances in which the methods perform well and those in which they perform poorly.

None of the estimates of the plant population abundance was equal to the true abundance. This does not mean that the methods performed poorly. If we were to redo the surveys we would get different numbers of detections and different estimates of abundance. There is randomness in what is seen and in the abundance estimates. In the chapters that follow, we develop methods to quantify the uncertainty that results from this randomness.

1.6   Outline of the book

We mentioned above that we show the underlying common features running through all abundance estimation methods. Our quantitative tool for doing it is the likelihood function. Nearly all the methods we consider can be formulated as special cases of a few general likelihood functions. Our focus throughout the book is very much on likelihood-based methods of inference.

We introduce likelihood functions and maximum likelihood estimation in Chapter 2. The remainder of the book is divided into three parts:

Simple Methods This part deals with methods that, with a few relatively straightforward exceptions, do not involve heterogeneity. It starts in Chapter 3 where we describe the basic conceptual and statistical objects that are used to construct the models and methods. In Chapters 4 to 8 we use these to build models for each of the "simple" abundance estimation methods. There is a general trend of increasing complexity across these chapters. Plot sampling methods (Chapter 4) are the simplest because they involve known detection probabilities. In Chapter 5 (removal methods) and Chapter 6 (mark-recapture), we consider methods that involve estimation of both detection (or capture) probability and abundance, but not heterogeneity. Chapter 7 covers distance sampling methods that involve estimation of detection probability and abundance, as well as a specific simple form of heterogeneity. Chapter 8 (nearest-neighbour and point-to-nearest-object methods) is essentially plot sampling in which plot size is a function of object density.

Advanced Methods In this part of the book we deal with spatial and temporal modelling of animal abundance, the issue of heterogeneity, and methods that combine various apparently different methods. In Chapter 9 we generalize some of the underlying conceptual and statistical objects introduced in Chapter 3. Chapter 10 deals with models that allow estimation of spatial and/or temporal distribution when detection of animals is certain within those parts of the survey region that are searched. Chapter 11 deals with ways of adapting the simple methods to accommodate heterogeneity. In Chapter 12 we cover methods that integrate methods of the preceding two chapters, allowing spatial and/or temporal distributions to be modelled when detection is uncertain even in those parts of the survey region that are searched. Several approaches that combine what at first appear to be unrelated methods are also developed. Some of the material in Chapters 11 and 12 is new and occasionally speculative, covering material beyond that in the published literature.

The last chapter in this part (Chapter 13) introduces methods for open populations. We have included it to suggest how methods for open populations are in many ways just the methods for closed populations with some added features, and to suggest how the general likelihood framework developed for the closed population case can be extended to deal with open populations. It does not cover open population methods in any detail.

Overview This part contains only one chapter, which is intended to give an overview of the strengths and weaknesses of each method and to indicate the situations in which each method is likely to be useful and those in which it is not. It is the only chapter in the book that contains no statistical or mathematical content.

We have tried to cater for readers with a range of statistical and mathematical competence. To this end, we have marked sections that involve substantial statistical or mathematical development using a different font and marginal notes.[1] Our hope is that this will help those who want to gain understanding of the methods without having to grasp the details of the statistical development – by allowing them to skip over the sections with substantial technical development. This is much easier to do in Part II than in Part III – by its nature the latter involves substantially more statistical development than the former. Part IV should be easily accessible to all readers.

At the start of each chapter, we highlight what we feel is the key idea underlying the methods of the chapter. We also give a point-form summary of the essential features of the model underlying the method and of the general form of likelihood function used in the chapter.

Solutions to selected exercises at the end of most chapters can be found by following links on the web page

http://www.ruwpa.st-and.ac.uk/estimating.abundance/

In time, we hope to put answers to all exercises on this web page.

1.7   R software

We have written a set of simulation and estimation functions covering many of the methods described in Part II. They are written in R and are available as the R library[2] WiSP (for "Wildlife survey Simulation Package") by following links from

---

[1]There is one exception: we do not mark these "technical" sections in Chapter 13 because the chapter is almost entirely composed of such sections.

[2]R is a language and environment for statistical computing and graphics. R is similar to the award-winning S system and is free. It provides a wide variety of statistical and graphical techniques.

http://www.ruwpa.st-and.ac.uk/estimating.abundance/

R itself is available via

http://www.r-project.org/

The library WiSP has been used frequently in Part II, to create plots, to simulate populations, estimate abundance, and so on. At least one version of the methods of each chapter of Part II is in WiSP; the methods of Part III are not. In time we hope to expand the library to include them.

Many of the R commands used to create exercises, plots, simulate populations, etc. for this book are available from the above site. There is also an introduction to the software and a tutorial to get you started with it.

The software is intended for learning and teaching rather than estimating abundance in real applications. There is a variety of much more comprehensive software packages available[3] for many of the methods; WiSP is not intended to replace any of them and does not have their functionality.

---

[3] A useful lists of links can be found at http://www.mbr.nbs.gov/software.html.

2

Using likelihood for estimation

Key idea: choose the most likely value as the estimate, given what was observed.

Key notation:

       N: population size (abundance)
       $\hat{N}$:   estimator of population size
       n: number of animals detected (sample size)
       p: probability of detecting an animal

2.1 An example problem

It is often easier to understand how abundance estimation methods work if we can check our estimates against the true population after estimating abundance, to see how well we did. This is impractical with real populations, so we will be using examples with artificial populations for illustration.

One such population, used repeatedly in this book, is the one introduced in Chapter 1. The data are actually from independent surveys by eight

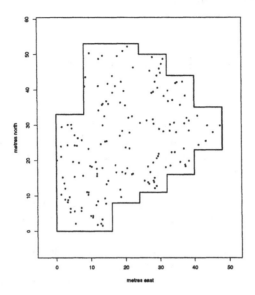

Figure 2.1. Example data, detected animals. Each dot represents a detected animal within the survey region. In all, $n = 162$ animals were detected.

different observers of a population of 250 groups (760 individuals) of golf tees, not plants, contrary to what we said in Chapter 1. The tees, of two colours, were placed in groups of between 1 and 8 in a survey region of 1,680 m^2, either exposed above the surrounding grass, or at least partly hidden by it. They were surveyed by the 1999 statistics honours class at the University of St Andrews,[1] Scotland, so while golf tees are clearly not animals (or plants), the survey was real, not simulated. We treat each group of golf tees as a single "animal", with size equal to the number of tees in the group; yellow tees are "male", green are "female"; tees exposed above the surrounding grass are classified as exposed ("exposure=1"), others as unexposed ("exposure=0").

Other populations presented later in the book were generated with the R library WiSP and only ever existed inside a computer. In all cases, we refer to them as animal populations, and to their members as animals.

Figure 2.1 shows the locations of the animals detected by at least one observer on a survey of our first example population. A total of $n = 162$ animals were seen, but an unknown number were missed. We would like to use what was seen to answer the question: How many animals are there?

---

[1] We are grateful to Miguel Bernal for making these data available to us. They were collected by him as part of a Masters project at the University of St Andrews. St Andrews is known as "the home of golf", so tees seemed an appropriate target object.

The method we use to answer this question throughout the book is "maximum likelihood estimation".

2.2   Maximum likelihood estimation

2.2.1   Known detection probability (p)

If we somehow knew that say half the animals had been seen, then there were exactly $162/\frac{1}{2} = 324$ animals in the population. In practice we never know the proportion of animals we saw on a survey. Suppose for this survey we knew that *on average* there was a 50% chance of detecting any animal. We could intuitively estimate the number of animals in the survey region (N) as before, by arguing that the 162 detected animals represent about 50% of the total number there, on average. That is $0.5 \times \hat{N} = 162$, or $\hat{N} = 324$.

The long-term average proportion of animals which would be detected, were we to replicate the survey very many times, is called the detection probability, and we refer to it as p. Our intuitive estimator can be written in terms of p as

$$\hat{N} = \frac{n}{p} \qquad (2.1)$$

This equation is the estimator; the number we get from it is the estimate.[2] When $p = 0.5$, the estimate of N is identical to that from the case above, when we knew that exactly 50% had been seen. The difference is that the estimate $\hat{N} = 324$ now has uncertainty associated with it, because the sample proportion detected on this occasion is uncertain (although its long-term average may be known). The uncertainty is expressed in terms of the variance of the estimator $\hat{N}$: the higher the variance, the higher the uncertainty. When the sample proportion is known, the uncertainty is zero and the variance is zero.

2.2.2   Unknown detection probability

In most real situations, we do not even know the expected proportion of detected animals (p). Clearly the estimator in Equation (2.1) depends critically on what we take p to be. For example if p had been 0.648 instead of 0.5, then our estimate of N would have been 250 rather than 324. In

---

[2] An estimator $\hat{N}$ is a *function* of the data (the function is $n/0.5$ in this case), and when we apply the function using the observed data, we get a *number* which is an estimate (324 in this case). We use the same notation, with "hats", for both estimators and estimates.

other words, in order to estimate N, we usually need to estimate another unknown quantity, the probability of detection, p.

The problem is that we don't know what the probability of detecting any one animal really is. All we know is that it is a probability, p, that lies somewhere between 0 and 1. The value of p depends on things like

(1) how easy the animals themselves are to detect (this will depend on their colour, size, etc.),

(2) how hard we look (the more person-hours we invest looking, the more we will detect), and

(3) environmental conditions (they may be less detectable in low light conditions, in certain habitats, etc.)

Much of the rest of the book is concerned with how one estimates N when p is unknown. Depending on how we go about looking, p is unlikely to be the same number for all animals in the target population. For the moment, however, let us suppose that we know p.

2.2.3   Basics of the method

Our purpose now is to give another rationale for the estimator $\hat{N} = n/p$. We will show that it gives us the most likely value of N, given what we observed. It is called the **maximum likelihood estimator** (or MLE) of N. The reason for showing this is not to complicate something that is really very simple, but rather to use this simple example to illustrate the principles and basic ideas of maximum likelihood estimation.

In more complicated situations, it is not always obvious how to estimate N. The maximum likelihood method then provides a general approach, and is the key to the methods of estimation discussed in this book. It can also provide a means of assessing the precision of our estimator of N. We know that the estimate, say $\hat{N} = 324$, is probably not exactly equal to N so it is useful to know for example that with 95% confidence N is somewhere between 292 and 362 animals.

We will often need to make a further assumption, which is not always applicable and not always easy to justify. This is the assumption that animals are detected or captured independently. In our example, this requires that the probability of detecting any one animal does not depend on which other animals were detected. If, for example, one animal flushes and causes others to flush and be seen, this assumption would not hold. Roughly speaking, one must be able to regard each detection as a separate experiment, like the toss of a coin, which is not influenced by how the coin might have landed on previous tosses. Fortunately methods can often be made robust (insensitive) to failures of this assumption.

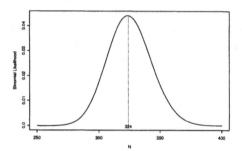

Figure 2.2. The binomial likelihood for N, given $n = 162$ and $p = 0.5$. The dotted line shows the location of the MLE, $\hat{N} = 324$.

Given that the detections occur independently and given that there are N animals in the survey region, it is possible to compute the (exact) probability that we will see n of them. This probability is given by the binomial distribution (see Section 2.2.4 below). For example, if $N = 305$ and $p = 0.5$, then the probability of detecting $n = 162$ is about 2.5%. It rises to about 4.5% when $N = 324$, and then falls to about 2.5% again when $N = 343$. Figure 2.2 is a plot of these probabilities over all N for which the probability is noticeably greater than zero. Based on our assumption of independence, and that $p = 0.5$, it shows the likelihood of the true N (which we don't know), given that $n = 162$ animals were observed.[3] The N at which this plot is a maximum is the maximum likelihood estimate of N, which is indicated by the dashed vertical line in Figure 2.2. The formula from which the plot was obtained is called the likelihood function.

Under our assumptions, the maximum of the likelihood function coincides with our intuitive estimator of N, i.e. the MLE is $\hat{N} = n/p = 324$.

Figure 2.2 illustrates the idea of maximum likelihood without any of the underlying mathematics. Once you understand the idea of maximum likelihood, it is often easier to find the MLE mathematically, using the formula for the likelihood function, than by plotting the likelihood.

2.2.4 Likelihood function

To construct the likelihood function, we need to formulate the estimation problem in statistical terms, and to do this we need some notation. We will call the whole region in which we want to estimate abundance, the **survey region**. (The region within the irregular border shown in Figure 2.1 is an example.) We call the

Technical
section
↓

---

[3]The figure should strictly be made up of points corresponding only to integer values cf N. We have drawn it as a continuous curve for ease of presentation.

number of animals in the survey region N, and the number which are detected, n. Our assumptions in this example can be stated formally as:

(1) The probability of detecting any animal in the survey region is $p = 0.5$.

(2) Detections are independent events.

With these assumptions, we can think of the survey as consisting of N independent trials, where each animal is a trial: a "successful" outcome is when the animal is detected, and a "failure" is when it is not. These are exactly the conditions underlying the derivation of the binomial distribution, so n is a binomial random variable, with parameters N (which we don't know) and $p = 0.5$. Its probability density function ("pdf" for short - which gives the probability of getting exactly n "successes" on the survey) is

$$f(n; N, p) \;\; = \;\; \binom{N}{n} p^n (1-p)^{N-n} \qquad (2.2)$$

where $\binom{N}{n}$ is the number of ways of choosing the n animals to appear in the sample from the N animals in the population. Now when we have done the survey, we know n (we observe it), and since we assume that $p = 0.5$, the only unknown quantity is N. Considered as a function of N in this way, Equation (2.2) is the likelihood function for N. We write it

$$L(N|n, p) \;\; = \;\; \binom{N}{n} p^n (1-p)^{N-n} \qquad (2.3)$$

We often omit all but the parameter(s) to be estimated from the argument list of likelihood functions, writing $L(N)$ instead of $L(N|n, p)$ for example. Although the right-hand side of this equation is the same as that in Equation (2.2), we plot and evaluate the likelihood as a function of N, whereas N is treated as a fixed constant in Equation (2.2). The N at which this likelihood function achieves its maximum value is the MLE of N.

We can maximize Equation (2.3) by finding the values of N for which the likelihood is increasing with N, i.e.:

$$\frac{L(N|n, p)}{L(N-1|n, p)} \;\; > \;\; 1 \qquad (2.4)$$

$$\Rightarrow \frac{N}{N-n}(1-p) > 1$$

$$\Rightarrow (1-p) > 1 - \frac{n}{N}$$

$$\Rightarrow N < \frac{n}{p}$$

The maximum of the likelihood function therefore is at the largest $\hat{N}$ such that $\hat{N} < n/p$. If we don't allow fractions of animals, this is the integer part of n/p, i.e. $\hat{N} = [n/p]$.

An alternative general approach to finding the MLE involves differentiating with respect to N to find the point at which the likelihood function has zero slope, and is therefore at a maximum (or minimum). Because likelihood functions usually involve products, which can be changed to sums by taking logs, it is almost always easier to find the MLE by maximizing the log-likelihood function, $\ell(N|n,p) = \ln\{L(N|n,p)\}$ rather than the likelihood function. (The maximum log-likelihood coincides with the maximum likelihood because the log-likelihood is a strictly increasing function of the likelihood.)

The derivatives of terms in the likelihood which do not contain the parameter of interest (N) are zero, so we need not consider them. Hence

$$\ell(N|n,p) \propto \ln\{N!\} - \ln\{(N-n)!\} + (N-n)\ln\{1-p\} \tag{2.5}$$

where "$\propto$" means "proportional to", and indicates that we have omitted terms that are constants.

We now use the approximation

$$\frac{d\ln\{N!\}}{dN} \approx \ln\{N\} \tag{2.6}$$

to find the MLE. Treating the approximation as exact, we get

$$\frac{d\ell}{dN} = \ln\{N\} - \ln\{N-n\} + \ln\{1-p\} \tag{2.7}$$

By setting this derivative to zero and solving for N, we get the maximum likelihood estimator

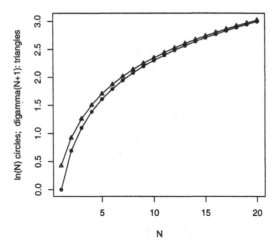

Figure 2.3. The digamma function, $\psi(N + 1)$ (dots) is not very close to $\ln\{N\}$ (triangles) for small N. Here the two are shown for $N \leq 20$.

$$\hat{N} \quad = \quad \frac{n}{p} \tag{2.8}$$

At this point we can discard the non-integer part of $\hat{N}$ if we don't want fractional animals.

Note that $\frac{d\ln\{N!\}}{dN}$ is the digamma function (sometimes written $\psi(N + 1)$; Kotz and Johnson, 1982), and while the approximation of Equation (2.6) is easier than the digamma function to deal with algebraically (which is why we use it frequently in this book), it can give misleading results, especially for small N. Figure 2.3 illustrates the difference between the approximate and exact results for $N \leq 20$.

For later use, we note that the variance of a binomial random variable n is $Np(1 - p)$. Since p is a known constant in this example, the variance of $\hat{N} = n/p$ is $Np(1 - p)/p^2$, and its standard error is $se[\hat{N}] = \sqrt{N(1 - p)/p}$. If $N = 324$ and $p = 0.5$, $se[\hat{N}] = 18$.

↑
Technical
section

2.3   Estimator uncertainty and confidence intervals

2.3.1   What are confidence intervals?

If the survey was repeated, it is very likely that a different number of animals would be seen. This would be true each time the survey was repeated

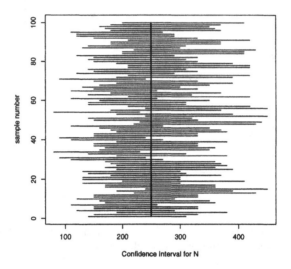

Figure 2.4. 95% confidence intervals (CIs) from 100 surveys of a population of $N = 250$ animals, with detection probability $p = 0.1$. Horizontal lines are CIs from individual surveys; the vertical line indicates the true N.

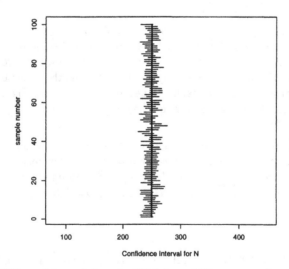

Figure 2.5. 95% confidence intervals (CIs) from 100 surveys of a population of $N = 250$ animals, with detection probability $p = 0.9$. Horizontal lines are CIs from individual surveys; the vertical line indicates the true N.

(providing observers do not remember what they saw last time). The number of animals detected, n, is a random variable; it varies from one survey occasion to another. Because it involves n, the estimator $\hat{N} = n/p$ is a random variable too. The more it varies between surveys, the less confidence we would have in our estimate of N from any survey. We are able to quantify this, and we do so using "confidence intervals".

A 95% confidence interval for N is just a *set* of estimates of N (as opposed to a single point estimate of N, like $\hat{N} = 324$), which would include the true N 95% of the time if the survey was repeated an infinite number of times. A confidence interval is also a random variable, because it is a function of (i.e. depends on) the data, n. Think of it as an interval which shifts in a random way from one survey occasion to another, but in such a way that it includes the true N 95% of the time. Figure 2.4 illustrates the situation. We can't plot results from an infinite number of surveys; the figure shows the 95% confidence intervals for N obtained from 100 surveys of a population of $N = 250$ animals when $p = 0.1$. The dark vertical line indicates the true N. Each horizontal line is a confidence interval obtained from a single survey. If we plotted confidence intervals for enough such samples, exactly 95% of the horizontal lines (confidence intervals) would touch (include) N. The confidence interval from any one survey has a 95% chance of including it. In reality we only have one survey and one confidence interval, and we don't know whether it is one that actually includes N or not, but we do know that it includes N with 95% probability.

Figure 2.5 shows the 95% confidence intervals for N obtained from 100 surveys of the same population of $N = 250$ animals, but when $p = 0.9$. The confidence intervals are much narrower than those in Figure 2.4 because there is much less uncertainty about N from a survey with $p = 0.9$ (in which 90% of the population is detected on one survey on average) than from a survey with $p = 0.1$ (in which 10% of the population is detected on one survey on average). In both cases however the 95% confidence intervals include N 95% of the time on average.

We refer to confidence intervals for N as CI[N], and we specify them via their lower and upper limits. These are sometimes expressed as $\hat{N}_{lower} = \hat{N} - d_L$ and $\hat{N}_{upper} = \hat{N} + d_U$, where d_L is the distance from $\hat{N}$ to the lower limit and d_U is the distance from $\hat{N}$ to the upper limit. For example, if $\hat{N}$ was a normal random variable (which it is not), you would probably recognize the form of the 95% confidence intervals as $\hat{N} - (1.96 \times se[\hat{N}])$ and $\hat{N} + (1.96 \times se[\hat{N}])$.[4] In this case the confidence interval is symmetric about $\hat{N}$ so that $d_L = d_U = (1.96 \times se[\hat{N}])$.

If, for a given confidence level, the confidence interval is narrow, we have estimated N with high precision; if it is wide, we have estimated N with low precision. The confidence level is most often taken to be 95%.

---

[4] $se[\hat{N}]$ is the standard error of $\hat{N}$ and is equal to $\sqrt{Var[\hat{N}]}$.

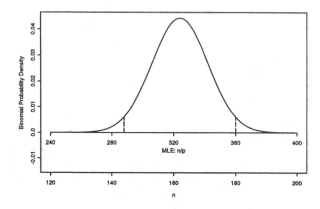

Figure 2.6. The distribution of n and of the MLE $\hat{N} = n/p$, when $N = 324$ and $p = 0.5$. The dashed lines indicate the 2.5 and 97.5 percentiles.

2.3.2   Constructing confidence intervals

Exact confidence intervals

How do we construct confidence intervals? Under our assumptions about the way the survey works, the probability that we count exactly n animals is given by the binomial distribution (see Section 2.2.4). For any given abundance, N, the binomial distribution tells us the proportion of time that we would count each of $n = 0$, $n = 1$, ..., $n = 162$, ..., $n = N$ animals on average (were the observers to conduct an infinite number of these surveys). We plot these proportions for $N = 24$ and n ranging from 120 to 200 in Figure 2.6. Because our estimator of N is just the random variable n divided by a known number $p = 0.5$, Figure 2.6 also gives the **"sampling distribution"** of our estimator, $\hat{N}$ (see the inner horizontal axis labelled "MLE: n/p" in the figure). This tells us the proportion of time we would get $\hat{N} = 120/0.5$, ..., $\hat{N} = 200/0.5$.

For any N, we can also calculate the proportion of time that we would count no more than $n = 162$ animals, by adding up the heights of the binomial distribution with parameters N and p for all integers $n \leq 162$. Similarly, we can calculate the proportion of time we would count at least $n = 162$ animals by adding up heights from $n = 162$ to the right. Figure 2.7 shows two binomial distributions. The one on the left is for $N = 292$. If this was the true N, we would observe an n of at least 162 about 2.5% of the time (the sum of heights to the right of 162 is about 2.5%). If the true N was 362, the distribution on the right would apply, and we would observe

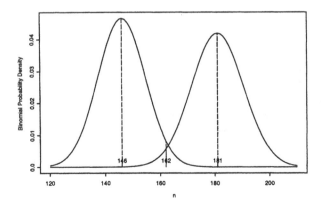

Figure 2.7. The distribution of n when $N = 292$ (on the left) and when $N = 362$ (on the right). The observed $n = 162$ is also shown. The area under the left curve to the right of 162 is about 2.5%, as is the area under the right curve to the left of 162.

an n of no more than 162 about 2.5% of the time (the sum of heights to the left of 162 is about 2.5%).[5]

The interval $(292, 362)$ is the 95% confidence interval for N. Note that if $N = 292$, we would estimate abundance to be equal to or greater than the survey estimate $\hat{N} = 324$ about 2.5% of the time, and if $N = 362$, we would estimate it to be 324 or less about 2.5% of the time (see Figure 2.8). By definition, $\hat{N}$ falls within the central 95% of its distribution 95% of the time. So even though we don't know the true distribution of $\hat{N}$, because we don't know N, if we construct confidence intervals by finding the two Ns whose binomial distributions have $\hat{N}$ as their lower and upper 2.5 percentile points, then this interval will include the true N 95% of the time.

Another way to think about confidence intervals is in terms of hypothesis tests. For any $N < 292$ and any $N > 362$, our observation of $n = 162$ would result in rejection at the 5% significance level of the null hypothesis that this was the true N against the two-sided alternative hypothesis that it was not. Conversely, the null hypothesis that any N in the interval $(292, 362)$ is the true N would not be rejected at the 5% level. The set of Ns in the 95% confidence interval are those that are supported by a two-sided hypotheses test of size 5%.

This is a particularly simple example because we know p and can easily evaluate the probability density function (pdf) of n for any N. Often we

---

[5]The reason we say "about 2.5%" is that there are no integer ns corresponding exactly to the 2.5 percentile point or the 97.5 percentile point for these distributions.

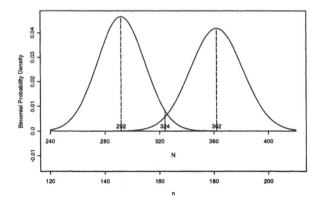

Figure 2.8. As Figure 2.7, but with the sampling distributions of $\hat{N}$ instead of the distributions of n, for $N = 292$ and $N = 362$. The MLE, $\hat{N} = 324$ is also shown.

will not be in this fortunate position, so we need other methods of getting confidence intervals. We outline three methods below.

2.4 Approximate confidence intervals

2.4.1 Asymptotic normality

A common way of estimating a 95% confidence interval is to approximate the distribution of the estimator ($\hat{N}$) with the normal distribution. Sometimes this approximation is good, sometimes not. It can be very poor when sample size is small and/or p is near 0 or 1. There are often better methods, but assumed normality is common and if you are familiar with it you may find the profile likelihood method below easier to understand by analogy. In our example, the binomial distributions shown in Figure 2.7 are close to normal distributions, so the method works well. From our sample, we estimate the standard error of $\hat{N}$ (the square root of its variance) to be 18 (see Section 2.2.4 above). Assuming that $\hat{N}$ is normally distributed, we estimate the 95% CI[N] to be ($\hat{N} - 36; \hat{N} + 36$) = (288, 360), to the nearest integer, and using a "continuity correction" to correct for the fact that the normal distribution is defined for all real numbers while the binomial is defined only for integers. (In this particular case use of $d_U = d_L = 1.96 \times \hat{se}[\hat{N}]$ and the continuity correction happen to have given an interval with $d_U = d_L = 2 \times \hat{se}[\hat{N}]$.) This is close to the exact interval of (292, 362).

Normal confidence intervals are by definition symmetrical about $\hat{N}$, but in our case the exact confidence intervals are not. The only information used to construct normal confidence intervals is the point estimate $(\hat{N})$ and an estimate of its variance. The approximate normal confidence interval does not incorporate any information about the shape of the distribution from which n and $\hat{N}$ arose.

The asymptotic normality of maximum likelihood estimators and functions of maximum likelihood estimators provides theoretical justification for assuming normality of $\hat{N}$ when sample size is large. As long as the likelihood function is reasonably "well behaved" and sample size is large enough, it turns out that the distribution of any MLE is close to normal. The larger the sample size, the better this approximation becomes, and the closer the expected value of the MLE is to the true value of the parameter (i.e. it is asymptotically unbiased). Its asymptotic[6] variance can also be calculated from the likelihood function. A summary of relevant results is given in Appendix C.

Confidence intervals constructed by assuming that $\hat{N}$ is lognormally distributed sometimes perform better than those constructed by assuming that $\hat{N}$ is normally distributed. Unlike normal-based CI's, they do not allow the lower bound of the CI to go below zero and they are not symmetric. See Burnham *et al.* (1987, p212) for details. For a $100(1-2\alpha)\%$ confidence interval, they have the form $(\hat{N}/d, \hat{N} \times d)$, with

$$d = \exp\left\{ z_\alpha \sqrt{\widehat{Var}[\ln(\hat{N})]} \right\} \tag{2.9}$$

where z_α is the upper α point of the standard normal distribution ($z_\alpha = z_{0.025} = 1.96$ for a 95% confidence interval), and

$$\widehat{Var}[\ln(\hat{N})] = \ln\left\{ 1 + \frac{\widehat{Var}[\hat{N}]}{\hat{N}^2} \right\} \tag{2.10}$$

2.4.2   Profile likelihood

The approximate normal confidence interval is based on the assumption that a known function of the data (i.e. the estimator $\hat{N}$) has an approximately normal distribution. The profile likelihood method is similar in that it is also based on a known function of the data having approximately a known distribution, but it uses a different function and a different approximating distribution (see below). Profile likelihood confidence intervals are in general closer to exact confidence intervals because they incorporate information about the shape of the sampling distribution of $\hat{N}$.

---

[6]That is, as n gets large.

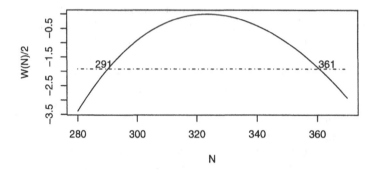

Figure 2.9. The difference between the (profile) log-likelihood function, $l(N)$, and its maximum, $l(\hat{N})$, for $280 \leq N \leq 370$, when $n = 162$ and $p = 0.5$.

When there is only one unknown parameter, as is the case here, the profile likelihood function is identical to the likelihood function. For the moment we restrict ourselves to this case, although it is when there is more than one parameter that the profile likelihood method is really useful. We deal with the more general case in Chapter 5 for the first time. The profile log-likelihood for N is shown in Figure 2.9,[7] together with the corresponding 95% confidence interval limits.

We plot the log-likelihood because the function whose approximate distribution is used to get profile likelihood confidence intervals involves the log-likelihood function (see below). The horizontal line is at -1.92: all Ns for which the curve is above the line are within the confidence interval. The Ns corresponding to the log-likelihood values closest to the horizontal line are shown next to the intersection of the two lines. These comprise the lower and upper bounds of the 95% confidence interval.

The profile likelihood 95% CI is (291, 361), which is closer to the exact CI than the normal confidence interval. Like the exact interval, the profile likelihood CI is not symmetric about $\hat{N}$.

Rationale for profile likelihood confidence intervals

Technical
section
↓

Let $L(N, \underline{\theta})$ be the likelihood function for the unknown parameters N and $\underline{\theta}$. We want to get a confidence interval for N, and $\underline{\theta}$ is a vector of "nuisance parameters"; unknown parameters that we are not really interested in, except insofar as they affect our CI for N. For example, p is a nuisance parameter when it is unknown, because we are really interested in N not p. If p was the only nuisance parameter, the shape of the profile likelihood for N is the

---

[7]The function is really defined only at integer N, but we show it as a continuous curve for convenience.

shape of its outline when viewed from a distant point perpendicular to the N axis, its profile. (We deal with this case in Chapter 5.) If there are no nuisance parameters (as in the example above), the profile likelihood for N is identical to the likelihood for N.

In general, the value of the profile likelihood at any N is the maximum of the likelihood function, with respect to the nuisance parameters, $\underline{\theta}$, at that N. The profile likelihood function for N is the set of these values for all N. We write it $L(N, \hat{\underline{\theta}}(N))$, where $\hat{\underline{\theta}}(N)$ is the value of $\underline{\theta}$ which maximizes the likelihood function at the given N. At $N = \hat{N}$ (the MLE of N), the value of the profile likelihood function is equal to the maximum of the likelihood function, $L(\hat{N}, \hat{\underline{\theta}})$, and standard large-sample likelihood results tell us that the likelihood ratio statistic

$$W(N) = 2\left[l(\hat{N}, \hat{\underline{\theta}}) - l(N, \hat{\underline{\theta}}(N))\right] \qquad (2.11)$$

has approximately a χ_1^2 distribution. (Here $l(\ )$ is the log of the likelihood function: $l(\ ) = \ln\{L(\ )\}$.) This approximate distribution of $W(N)$ holds better at small sample sizes than the approximately normal distribution of $\hat{N}$, which is why profile likelihood confidence intervals are generally closer to exact confidence intervals.

Using this result, the profile likelihood 95% CI for N is the range of Ns for which $W(N) \leq 3.84$ (the 95% point of the χ_1^2 distribution).[8] Equivalently, it is the range of Ns for which the profile log-likelihood is within $3.84/2 = 1.92$ of its maximum value.

↑
Technical
section

2.4.3 Bootstrap

In this book we frequently use simple "**bootstrap**" methods to estimate confidence intervals and variances. See Manly (1997) for a brief introduction to bootstrap methods in biology, and Davison and Hinkley (1997) and Efron and Tibshirani (1993) for more thorough introductions to bootstrap methods in general. Their advantage as a teaching tool is that they are quite easy to understand and are applicable to a wide range of problems.

In the context of our current example, bootstrapping consists of building up the sampling distribution of $\hat{N}$ by doing a large number of surveys inside the computer, and counting the proportion of times we get each $\hat{N}$. There are many ways of doing this, but the two basic methods are

---

[8]Note that $3.84 = 1.96^2$; this relationship between the confidence limits for the χ_1^2 and standard normal distributions follows from the fact that the square of a standard normal random variable has a χ_1^2 distribution.

the "**parametric bootstrap** , in which we use a statistical model to do the resampling, and

the "**nonparametric bootstrap**" , in which we resample the observations.

Parametric bootstrap

For the parametric bootstrap, we conduct many surveys in the computer using the statistical model we used to estimate N. In the context of the current example, we generate a random n using the binomial probability density function (Equation (2.2)). We know that we should use $p = 0.5$ (by assumption) but we don't know the correct N to use, so the computer surveys will not be quite like repeated real surveys of the population. As we don't know N, we do the next best thing and use our estimate $\hat{N} = 324$ in its place. We use the assumed parameter $p = 0.5$ and the estimated parameter $\hat{N} = 324$ to simulate the surveys in the computer, hence the "parametric" in the name of the method. This works well in many cases. Figures 2.10 to 2.12 illustrate how we build up the distribution of $\hat{N} = n/p$ by generating n for each of a large number of computer surveys.

The method we use in this book to find confidence intervals from bootstrapping is called the "percentile method". It is simple, does not constrain the interval to be symmetric about the estimate (unlike normal-based intervals), and often works well. A percentile method confidence interval is constructed as follows. Once we have run enough surveys in the computer, we get the 95% confidence interval by finding the points on the $\hat{N}$-axis of the plot below which 2.5% of the computer-generated $\hat{N}$s fall, and above which 2.5% of the computer-generated $\hat{N}$s fall. These points are shown by the dashed lines in Figures 2.10 to 2.12. The bootstrap CI$[N] = (294; 348)$ from our 79 bootstrap resamples is not very close to the correct interval. With 199 resamples, it is better: CI$[N] = (290; 356)$, and with 10,000 resamples, we get the interval: CI$[N] = (288; 360)$. While this is no better than the confidence interval obtained by assuming normality in this example, the percentile method CI for N is generally more robust, particularly with small sample size and skew distributions. (When the distribution of $\hat{N}$ is skew, the percentile method is counter-intuitive because it seems to get the asymmetry the wrong way round, but see Davison and Hinkley (1997, pp202-3) for a rationale of the method in this case.)

Note that we can also easily calculate an estimate of the variance of $\hat{N}$ from our bootstrap samples, by calculating the variance of the many $\hat{N}$s we generated in the computer.

It may be difficult to see the merits of the bootstrap method from this example, because the true distribution of $\hat{N}$ is known. It is in more complicated situations that bootstrap methods are really useful. These arise later in the book.

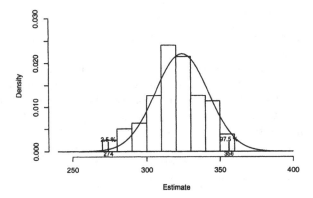

Figure 2.10. A bootstrap distribution of n and $\hat{N} = n/p$, when $N = 324$ and $p = 0.5$, from 79 resamples. The solid line shows the true distribution. The estimated 95% confidence limits using the percentile method are indicated by short vertical lines.

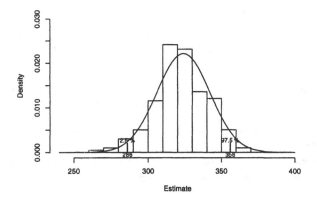

Figure 2.11. As Figure 2.10, but with 199 bootstrap resamples.

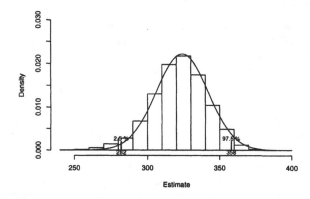

Figure 2.12. As Figure 2.10, but with 10,000 bootstrap resamples.

Nonparametric bootstrap

The nonparametric bootstrap works in much the same way as the parametric bootstrap, with one important difference: it does not use a statistical model when simulating surveys in the computer, but resamples the observations with replacement. With our simple example, the two methods are equivalent: for the parametric bootstrap, we generate a random deviate from the binomial distribution with parameters $N = 324$ and $p = 0.5$; for the nonparametric bootstrap we sample 324 values with replacement from $n = 162$ captured animals and $\hat{N} - n = 162$ animals not caught. Hence, we again get a deviate from the binomial distribution with parameters $N = 324$ and $p = 0.5$. In general, the methods differ.

2.5   Summary

This chapter introduced the concepts of the likelihood function and maximum likelihood estimation.

The likelihood function is a function which tells us the likelihood of any unknown "truth" (N in our example), given the observed data (n in our example).

The maximum likelihood estimate (MLE) is the most likely value of unknown "truth", given what was observed.

To evaluate the likelihood function, and find the MLE, we need to construct a statistical model describing how the data were generated. The model is built on the basis of assumptions which we believe, or are prepared to accept. In the example we looked at, we assumed that:

(1) Animals are detected independently of one another.

(2) The probability of detecting any one animal is $p = 0.5$.

From these assumptions, it follows that n has a binomial distribution, and the MLE of N is $\hat{N} = n/p$. Using this model, we illustrated a number of methods of obtaining confidence intervals for N.

Estimates and conclusions are only as good as the assumptions on which they are based, and in the case we looked at, we really had no basis for making the assumption that $p = 0.5$. To make it worse, the MLE is very sensitive to changes in p (it is inversely proportional to p). Estimates based on such a strong and arbitrary assumption should be regarded with extreme scepticism. We made the assumption here in order to illustrate the method of maximum likelihood as simply as possible. In most applications p is not known, and the majority of animal abundance estimation methods involve estimation of both N and p. We deal with this more complicated situation in later chapters.

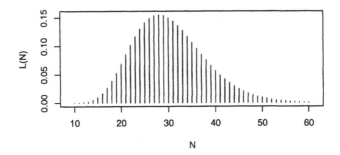

Figure 2.13. Binomial likelihood function for N, given p and $n = 10$.

2.6   Exercises

Exercise 2.1   For the example in this chapter, write down the MLE, and an associated 95% confidence interval, for abundance N, if $n = 162$ animals were detected and $p = 0.65$.

Exercise 2.2   Write down the likelihood function for p, using the same assumptions as were used to obtain a likelihood for N in this chapter, but with the assumption that p is known replaced by the assumption that N is known. Show that if N is 250 and 64 animals are detected, the MLE for p is $\hat{p} = 0.256$.

Exercise 2.3   Figure 2.13 shows the binomial likelihood function for N from a survey in which 10 animals were detected.

(a) What is the MLE of N?

(b) Calculate a normal-based confidence interval for N.

(c) Calculate a profile-likelihood confidence interval for N.

(d) Which of the confidence intervals in (b) and (c) is the more reliable, and why?

Exercise 2.4   Figure 2.14 shows the locations of animals in the example population that were detected by observer 1. Assuming that detections are independent and $p = 0.25$, do the following:

(a) Use these data to estimate the number of animals in the region, N, by maximum likelihood.

(b) Write down an expression for the variance of the MLE $\hat{N}$ obtained in (a), and hence estimate the variance of $\hat{N}$.

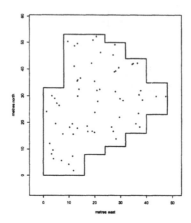

Figure 2.14. Example data, animals detected by observer number 1.

(c) Assuming that $\hat{N}$ is normally distributed, estimate a 95% confidence interval for N.

(d) [9]Treating the approximation $\frac{d\ln(N!)}{dN} = \ln(N)$ as exact, show that the Fisher information (which is a scalar in this case) is

$$I_{\hat{N}} = -E\left[\frac{d^2 l(N)}{dN^2}\bigg|_{N=\hat{N}}\right] = \frac{p^2}{n(1-p)} \qquad (2.12)$$

(e) [10]Hence show that the information matrix-based estimate of $Var[\hat{N}]$ is $\frac{n(1-p)}{p^2}$.

Exercise 2.5 N animals are located in a survey region of size A.

(a) If a particular animal is equally likely to be anywhere in the region, what is the probability that it is inside a "covered region" which has surface area a? (The "covered region" is that part of the survey region that is searched.)

(b) If all animals position themselves independently of all other animals, and all are equally likely to be anywhere in the survey region, what is the probability that there are n animals in the covered region? (Hint: think of it as a "success" if an animal falls in the covered region.)

---

[9]You need results from Appendix C to do this question.
[10]You need results from Appendix C to do this question.

(c) Hence write down the likelihood for N, given that n animals fell in the covered region.

Exercise 2.6

(a) Use the results from the previous question to obtain an MLE for N if 10% of the survey region is searched, and found to contain 10 animals.

(b) Write down an estimate of the variance of this estimator.

(c) Animals tend not to distribute themselves independently of one another. If they cluster together, would you expect the variance estimate obtained in (b) to be positively or negatively biased, and why?

Exercise 2.7   Given that the first observer (Camilla) on a survey of the example dataset saw $n_1 = 64$ of the animals in the survey region, and that all animals are equally detectable, do the following:

(a) Write down an expression for the probability that the second observer (Ian) sees m_2 of the animals seen by Camilla.

(b) Treat the expression in (a) as a likelihood for the probability, p of Ian seeing an animal. Given that Ian saw $m_2 = 34$ of the animals Camilla saw, calculate the maximum likelihood estimate, $\hat{p}$.

(c) [11]Show that the Fisher information (which is a scalar not a matrix in this case) is

$$I_{\hat{p}} = -E\left[\frac{d^2 l(p)}{dp^2}\bigg|_{p=\hat{p}}\right] = \frac{m_2}{p(1-p)} \qquad (2.13)$$

(d) Hence estimate a normal-based confidence interval for p.

(e) Calculate a profile-likelihood confidence interval for N.

(f) Compare the confidence intervals in (d) and (e) and comment on their similarity and the reasons for this.

---

[11]You need results from Appendix C to do this question.

Part II

Simple Methods

3
Building blocks

Key ideas:

State model:	statistical model describing the state of the population: the distribution and characteristics of animals in it.
Observation model:	statistical model describing how animals are detected or captured, given the searched region and their locations and characteristics.

3.1 State and observation models

Throughout this book we formulate the problem of estimating animal abundance in terms of state models and observation models. This gives a common framework and allows many different estimation methods to be viewed as variants on the same basic theme. It also provides a natural link to more complicated methods for open populations, which we discuss briefly in this book, but which are crucial for the management of wildlife populations.

The likelihood function provides the link between what we observe and what we want to estimate. At the heart of likelihood functions for estimating animal abundance are the state and observation models. Given a

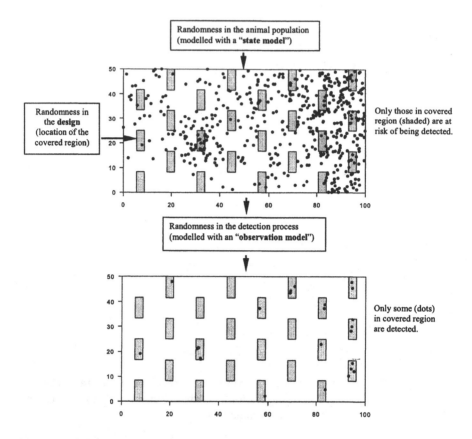

Figure 3.1. Schematic diagram illustrating the sources of randomness determining what is observed on a survey (bottom plot). The state model captures randomness in the population, the survey design determines which animals are at risk of detection, and the observation model captures randomness in detection within the covered region.

searched region, the state and observation models describe how the observed data were generated (see Figure 3.1).

The processes that led to the data involve both characteristics of the animals (where they are, how big they are, etc.), and how animals are observed, given their characteristics. We try to capture the former in the **state model**. It describes statistically the mechanisms determining the distribution of the animals in space, etc. We describe the latter with what we call the **observation model**. It quantifies probabilities of detection or capture, given the animal characteristics and the nature of the survey.

In this chapter we introduce both models in general terms. The technical details of general observation models for the methods of Part II are given in Appendix B. In each of the following chapters, we define the specific

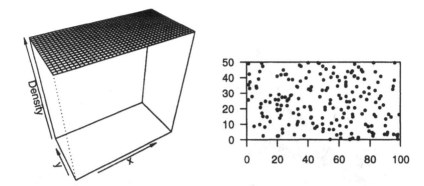

Figure 3.2. The mean of a uniform spatial state model (left) governing the distribution of $N = 200$ animals in a 50 km×100 km survey region. The height of the surface represents the expected animal density, which is the same throughout the survey region. A realization of the model (right) shows the actual locations of animals on this occasion.

state and/or observation models used to construct the likelihood functions used in the chapter.

3.1.1   State models

The state model is a statistical representation of the relevant biological, demographic and other processes which govern the distribution of animals with respect to location, sex, size, etc. We call the actual distribution at any time a "realization" of the state model. For example, Figure 3.2 shows the constant mean density associated with a uniform spatial state model, and the locations of $N = 200$ animals for one realization of this hypothetical population. Other realizations of the same state model would result in different animal locations.

In reality state models are seldom this simple. The mean density of a more complicated spatial state model, together with a realization of it, is shown in Figure 3.3. In this hypothetical state model, there is a steep density gradient in the east and local high-density patches scattered about the region, with one particularly high density patch in the centre west of the region. The high-density regions might be associated with regions of preferred habitat.

In the case of the St Andrews example dataset, the spatial state model comprised two uniform distributions – one for animals in the higher density southern stratum, the other for animals in the northern stratum. The realization of this state model (i.e. the locations of all animals in the survey region – detected and undetected) is shown in Figure 3.4.

A different realization of the same state model would lead to different animal locations; the average density of locations over an infinite number

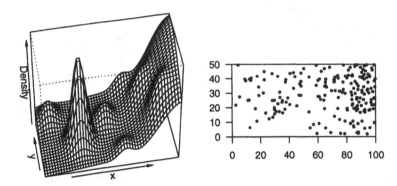

Figure 3.3. The mean of a more complicated spatial state model (left) and a realization of it with $N = 200$ animals. The axes are distance north (y) and distance east (x). See text for a fuller description of the plot.

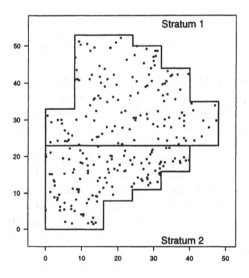

Figure 3.4. Example data, comprising a realization of two uniform spatial state models, one with $N = 130$ for animals in the northern stratum of area 1,040 m^2 (north of 23 m north), the other with $N = 120$ for animals in the southern stratum of area 640 m^2.

of realizations is the mean of the state model. With a uniform state model, this average is a flat horizontal surface – corresponding to equal average density everywhere the uniform model applies.

Definition of a simple uniform state model

We use the uniformly-distributed population shown in Figure 3.2 for illustration. Each animal's location is defined by its distance east (u, say) and distance north (v, say). A uniform spatial state model for a single animal can be described statistically as the product of uniform probability density functions for u and for v, where in our case 0 m$\leq u \leq$ 100 m and 0 m$\leq v \leq$ 50 m:

$$\pi(u,v) = \pi(u) \times \pi(v) = \frac{1}{100} \times \frac{1}{50} = \frac{1}{A} \quad (3.1)$$

where $A = 5,000$ km^2 is the area of the survey region.

We use the notation $\pi(\ )$ for probability density functions (pdf's) associated with state models. In this example, $\pi(u)$ is the pdf for distance east, and $\pi(v)$ is the pdf for distance north.

If, within this 50 m × 100 m survey region, animals are distributed independently according to Equation (3.1), then the joint pdf of the locations of all $N = 200$ animals in the survey region is just the product of the pdfs for each animal:

$$\pi((u_1;v_1),\ldots,(u_{200};v_{200})) = \prod_{i=1}^{200} \pi(u_i) \times \pi(v_i)$$
$$= \left(\frac{1}{A}\right)^{200} \quad (3.2)$$

where $(u_1;v_1),\ldots,(u_{200};v_{200})$ are locations of the 200 animals in the population.

The statistical definition of the density surface shown in Figure 3.3 is much more complicated and we don't show it here. If we assume that animals are distributed identically and independently of one another, then the complication lies only in specifying mathematically the shape of the mean surface shown in the figure. Estimating such a complicated surface reliably from limited data is a difficult problem. Not surprisingly, therefore, most state models used in practice are relatively simple. We deal with some approaches for estimating complex spatial state models in Chapter 11.

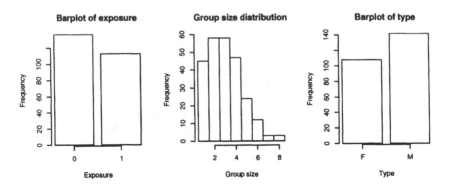

Figure 3.5. St Andrews example data, distribution of exposure level, group size, and type (M or F).

Non-spatial components of state models

Spatial state models are not the only relevant ones. Conceptually there are state models underlying the age distribution, sex distribution, size distribution, etc. in a population. For example, each animal in the St Andrews example dataset has three characteristics associated with it in addition to its location. They are its size, sex and exposure. The distribution of these three characteristics in the population is shown in Figure 3.5. These are realizations of state models for size, sex, and exposure. A plot of the animal characteristics together with their location is shown in Figure 3.6.

Because animal populations are dynamic, the state model should depend on time unless we are dealing only with a "snapshot" of the population at a single point in time. In Part II of this book we restrict ourselves to particularly simple snapshots, or to variables which do not change over the course of the surveys, and to "simple" abundance estimation methods. We define "simple" methods to be those which include at most a uniform spatial state model, the state model is treated as *known*, and only one realization of the state model is involved. In Part III of the book, which covers what we call "advanced" methods, we consider situations in which the state models are not completely known, and may include non-spatial components.

Note that although we do not include state models for factors like size or sex in the state models used for estimation in Part II, we do sometimes include these variables in the observation model (below) because detection/capture probability may depend on them.

The realization of the state model divides the N animals in the population into homogeneous sub-populations, so that there are $N(\underline{x})$ animals with animal-level variables $\underline{x}$. For example, if sex was the only relevant animal-level variable, $\underline{x}$ would indicate sex ($\underline{x} = 0$ for males, $\underline{x} = 1$ for

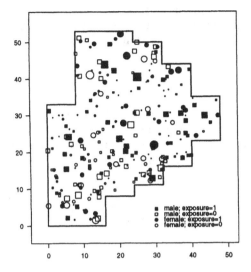

Figure 3.6. St Andrews example data, realization of the full state model. Symbol size is proportional to animal size.

females, say) and the population would be split into two sub-populations: $N(0)$ male animals and $N(1)$ female animals $(N(0) + N(1) = N)$.

3.1.2 Survey design and observation models

Observation models are statistical models describing how animals are detected or captured, given the variables associated with them. Only animals in the covered region can be detected so they apply only within the covered region. The covered region is usually chosen using some set of rules involving some randomness, called a survey design. The surveyor(s) control the survey design, including which observers/traps, and how many, are used, and which of a predefined set of survey units are included in the survey(s). The detection model describes the probabilistic process by which animals in these units are detected or captured.

Certain detection

The simplest possible observation model is for a single-survey scenario in which all animals in the covered region (the region that is searched) are detected with certainty, i.e., with probability, $p = 1$. We looked at this sort of scenario in the very first example of Chapter 1. Suppose that there are N_c animals in the covered region. There is no uncertainty about the number n that will be seen; the only possible outcome is that $n = N_c$. All uncertainty in estimated abundance comes from the fact that the whole survey region was not covered. If the whole population is in the covered region then

$\hat{N} = n$; if half the population is in the covered region, then $\hat{N} = 2n$, and so on. The state model specifies the locations of the animals and in this simple case it is the state model that is the source of all uncertainty in estimated abundance.

Detection with equal probability

The simplest possible observation model with uncertain detection is one in which all animals are detected independently with the same probability, $p < 1$. If there are N_c animals in the covered region, the probability that n are detected is given by the binomial pdf evaluated at n:

$$p(n) \quad = \quad \binom{N_c}{n} p^n (1 - p)^{N_c - n} \tag{3.3}$$

The number of animals in the covered region is now uncertain; we can estimate it by

$$\hat{N}_c \quad = \quad \frac{n}{p} \tag{3.4}$$

Because n is a random variable, $\hat{N}_c$ is a random variable. If the whole survey region was covered, then $N_c = N$ and $\hat{N} = \hat{N}_c$, so that all uncertainty comes from the observation model.

If the covered region is less than the survey region, then there is also uncertainty about what fraction of the population was in the covered region. Suppose that on average, a fraction π_c of the population is in the covered region. There is still uncertainty about the exact fraction in the covered region on any one realization, but we can use π_c to estimate N by

$$\hat{N} \quad = \quad \frac{\hat{N}_c}{\pi_c} \quad = \quad \frac{n}{p \times \pi_c} \tag{3.5}$$

The uncertainty in $\hat{N}$ contains uncertainty due to both the observation model (n is random) and the state model (the fraction of the population in the survey region is random).

Animal-level variables: "heterogeneity"

We have already covered the simplest of observation models in Chapter 2. With this model, all animals are equally likely to be detected (with probability p), and they are detected independently of one another. Figure 3.6 suggests that this model would not be adequate for our example data. From a comparison of detected vs. undetected animals (Figure 3.7), you might

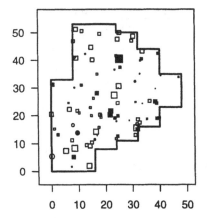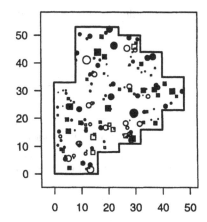

Figure 3.7. St Andrews example data: Characteristics of undetected animals (left plot) compared with detected animals (right plot). Symbol size is proportional to animal size; circles are females, squares are males; solid symbols are "exposed" animals, empty symbols are not.

guess that females (circles) are more detectable than males, exposed animals (solid symbol) are more detectable than unexposed animals, and it is not clear whether larger animals are any more detectable than smaller animals.

We cover the detail of how we can model detection probability below, but for the moment note that detection probability, p, is clearly not the same for all animals in the population. We refer to the presence of different ps in a population as "heterogeneity", and call this sort of population "heterogeneous". Nearly all animal abundance estimation methods give biased estimates of abundance if heterogeneity is ignored entirely when it is present, although some methods are much more sensitive to this than others. You can see why this is, without any statistical development, by thinking about an extreme case. Say a population consisted of only two types of animals, one of which you were virtually certain to detect, while the other was virtually impossible to detect. Unless you model them as two separate populations in some way, your survey will be little different from a survey of the detectable animals and your abundance estimate will tend to be very close to the abundance of the detectable animals alone.

The message here is that in general we need models which allow detection probability to depend on variables associated with the animals themselves, such as size or sex.

Before we leave this subsection, a brief note on animals that occur in groups. If it is the group that is the effective sampling unit (i.e. it is the group that is detected), we can regard the group as the "animal" and treat its size as an animal-level variable. If we do this, methods for estimating

Table 3.1. Numbers of animals detected by each observer

	Observer number							
	1	2	3	4	5	6	7	8
Number of detections	64	67	73	59	69	58	63	93

animal abundance for ungrouped populations can be used to estimate group abundance. Estimates of individual abundance in this case can be obtained by a variety of methods, including estimation by way of a state model for group size, or empirical estimation of mean group size.

Survey-level variables

Table 3.1 shows the number of animals detected by each of the 8 observers in surveys of our example population. Not surprisingly, some observers seem better than others i.e. detection probability may depend on observer. Detection probability also depends on the number of observers used. In short, detection probability depends on the characteristics of the survey itself (which observers were used, how many, etc.), in addition to depending on things associated with the animals (size, sex, etc.).

A critical difference between survey-level and animal-level variables is that we know the values of survey-level variables throughout the survey region, whereas we know the animal-level variables only for those animals we observe.

Detection with heterogeneity

Since detection probability can depend on things associated with the animals (size, sex, for example) as well as things associated with the survey itself (number and skill of observers, for example), a general model for the detection process may include both types of variable. If we call the animal-level variables $\underline{x}$, and the survey-level variables $\underline{l}$, then the detection probability is some function of the two:

$$p(\underline{x}, \underline{l}) \quad = \quad \text{Pr(detect animal of "type" } \underline{x}, \text{ using survey effort } \underline{l}) \quad (3.6)$$

Appendix B contains a detailed statistical formulation for observation models with heterogeneity.

3.2 Design and model

For all conventional survey designs, if we know what parts of the survey region were covered, the design itself (the set of rules used to choose the

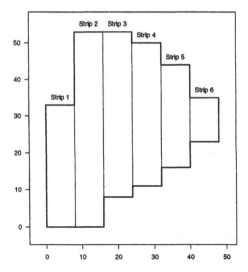

Figure 3.8. St Andrews example dataset design units: strips.

covered region) contributes nothing to the likelihood function. This is in stark contrast to the type of inference taught in sampling theory courses and books, in which the design is the only basis for inference.

In this section we briefly define the survey design, contrast design-based and model-based inference, and discuss the relevance of design to the model-based approach for inference in broad terms.

3.2.1   What is a survey design?

In our context, a survey design is a probabilistic rule for sampling from a set of predefined spatial units comprising the survey region. There are very many types of design, and a large literature on the subject. "Simple random sampling" (srs) is one of the simplest designs, and we use it for illustration here.

The survey region of our example data can be divided into a set of strip transects (Figure 3.8). Having decided how many of the strips are to be sampled, srs involves selecting them with equal probability and without replacement (i.e., so that every strip is equally likely to be sampled, and each strip can be selected at most once).

3.2.2   Design- vs. model-based inference

Note that the survey design introduces a new sort of randomness. Up to now we have spoken of (1) randomness in the location, size, sex, etc. of animals, and (2) randomness in whether or not a particular animal is detected. We try to capture (1) in our state model, and (2) in our observation model. The survey design introduces (3) randomness in whether or not a design unit (strip in our case) is searched. The nice thing about (3) is that we don't need to model it; we know *exactly* what it is and how it works, because we created it. If we could base our inferences about N (i.e. estimate N and an associated CI) on (3) alone, we would have no need for a state model, and no need for an observation model. This would be reassuring because we don't know the true state model or the true observation model, so we have to make some assumptions about them if we are to use them and, as we saw in Chapter 2, inferences can be misleading if our assumptions are wrong.

Inference based on the randomness introduced by the design alone is called design-based inference. It is rarely the case with wildlife surveys that no assumptions about the observation model are required, because it is rarely the case that we know p – unless detection within the sampled units is certain. While we can sometimes replace the state model with the survey design for inference, a purely design-based approach to estimating N in wildlife surveys is seldom possible. This is probably a large part of the reason that the wildlife survey literature overlaps so little with the conventional design-based sampling theory literature, and why you seldom see much on wildlife survey methods in sampling theory textbooks; an exception is Thompson (1992).

A simple example: point estimation

We deal with purely design-based inference in more detail in Chapter 4. For the moment, suppose that we can dispense with the observation model because we know that $p = 1$. We take a srs of four strips and count 132 animals in all. We want to estimate N, the total number of animals in the six strips.

With a likelihood-based approach (model-based inference), we estimate the average density over the survey region on the basis of our assumptions about the state model, given the observed counts in the covered region. For example, with a uniform state model we assume that, on average, the number of animals per unit area is the same in all strips. Since the combined surface area of strips 1, 3, 5 and 6 is 944 m^2, the estimated density is $132/944 \approx 0.14$ animals per m^2. The total survey area is 1,680 m^2, so under the assumption that average density is the same everywhere, we estimate there to be $1,680 \times 132/944 = 234.9$ animals in total (or 235 to the nearest integer). This does not depend in any way on *how* we chose strips 1, 3, 5 and 6 (i.e., it does not depend on the survey design).

Design-based estimation, by contrast, depends entirely on how the covered region was chosen. You can think of it as involving estimation of the average density *over all possible samples*, on the basis of the design, given the observed counts in the covered region for this sample. Because all strips are equally likely to be chosen, and 4 out of 6 are chosen, the total count in the chosen strips will average $\frac{4}{6}$ths of N if all possible simple random samples of size 4 are enumerated. Given a count of 132 in the chosen strips, we therefore estimate N to be $132/\frac{4}{6} = 198$. Note that we did not make any assumptions about the state model (distribution of animals between the strips) to get this estimator: our sample will include $\frac{4}{6}$ths of N, *on average* over all possible samples, no matter whether all N animals fall in strip 1, they are evenly distributed across strips, or they have any other distribution.

A simple example: interval estimation

Our discussion of design-based inference so far has covered only point estimation of N; we have not mentioned how confidence intervals can be obtained. Before we summarise the pros and cons of design-based vs. model-based inference, we say a few words about interval estimation. In the case of design-based estimation, the true distribution of $\hat{N}$ *does* depend on the distribution of animals across strips. For example, the variance will be higher if all animals are in one strip than if they are spread evenly over strips. While the variance of $\hat{N}$ can be unbiasedly estimated without any assumptions, confidence intervals can't. So at this point, design-based inference loses its entirely assumption-free nature; $\hat{N}$ is usually assumed to be normally distributed for the purposes of estimating confidence intervals. When sample size (number of strips in our example) is large, the Central Limit Theorem tells us that this assumption is good; when sample size is small, the assumption is likely to be violated.

Estimation of N, its variance, and confidence intervals are all model-based in the case of estimation purely via the likelihood function. As such, they may all be biased to some extent if the assumed model is inappropriate.

3.2.3   Can we tell if the model is wrong?

Yes, to some extent. For example, does the distribution of animals in the sample shown in Figure 3.9 look uniformly distributed? That is, does it look consistent with a uniform spatial state model?

What about the distribution of animals in the sample shown in Figure 3.10?

Survey data usually (not always) contain information that enables us to detect inappropriate models.[1]. In this case we have some basis for choosing among the contending models.

---

[1] Figure 3.9 is a sample from a population of 300 animals generated from the uniform spatial state model of Figure 3.2 Figure 3.10 is a sample from a population of 300 animals generated from the nonuniform spatial state model of Figure 3.3.

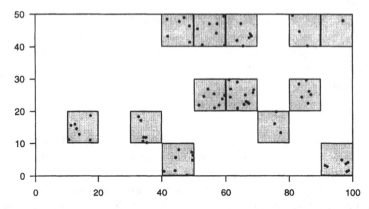

Figure 3.9. Plot sample 1. The shaded region was searched; dots are detected animals.

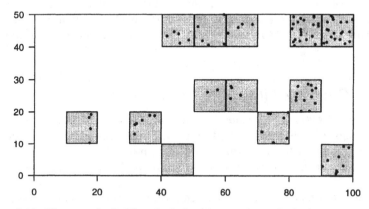

Figure 3.10. Plot sample 2. The shaded region was searched; dots are detected animals.

Model selection is a central part of model-based inference. It is a large subject in its own right and unfortunately we do not have space to go into it in any detail in this book. We consider one form of model selection briefly in the context of estimation for removal method, mark-recapture and line transect sampling in Part II.

There are many criteria and ways of choosing between competing statistical models for a survey dataset. We consider only Akaike's Information Criterion (AIC: Akaike, 1973). Given the maximum of the likelihood function, the AIC is easy to calculate, versatile, and conceptually fairly straightforward.

Recall that we maximize the likelihood function to get the MLE, and that the likelihood function is based on a statistical model.

- Let L be the value of the likelihood at its maximum.

- Let q be the number of unknown parameters in the likelihood (there is only one in the binomial likelihood we have looked at so far, namely N, so $q = 1$ in this case).

The AIC is defined as

$$\text{AIC} \;=\; -2\ln(L) + 2q \tag{3.7}$$

The first term, $-2\ln(L)$, gets smaller as the fit to the data improves (and the likelihood, L, increases). Now as you add more parameters to the model, it gets more and more flexible, and is able to fit the data better and better. On the basis of goodness of fit, you'd want to choose the model with the smallest first term. But the second term, $2q$, gets bigger and bigger as more and more parameters are added to the model. The second term is a penalty for having to estimate more and more parameters from the same amount of data. Choosing the model with the smallest AIC is a compromise between goodness of fit, and having to estimate many parameters. The criterion is applied by choosing the model with the smallest AIC among all the models under consideration.

3.2.4   Design-based vs. Model-based: pros and cons

(1) *Design-based point estimates can be assumption-free.* The biggest advantage of design-based inference over model-based inference is that the point estimate of N involves fewer assumptions than model-based inference. This is an advantage because we never know the true model(s), and inferences can be misleading if our assumptions about the model(s) are far off the mark.

(2) *Purely design-based inference is seldom possible* with wildlife surveys. Whenever the observation model is not completely known (which is

the case for most methods), some model-based inference is necessary to estimate p, and hence N.

(3) *Model-based inference allows us to estimate spatial distributions*, sex distribution, age distribution, size distribution, etc. By estimating the parameters of a spatial state model, we can readily draw inferences about which parts of the survey region have high densities of animals. While this can be done to some extent using design-based methods (by stratification for example), the capability is limited. Model-based methods allow more plausible smooth spatial models to be fitted and in general require fewer parameters to model spatial trend than design-based methods. Model-based methods readily extend to the estimation of state models for non-spatial variables like size, sex, etc.

(4) *Design-based variance estimates can be assumption-free.* Variance is usually estimated with respect to the survey design, which is known. Model-based variance estimates (like those from a parametric boot-strap, for example) are based on the assumptions of the model and are subject to bias if the model is inappropriate. In particular, state models often involve the assumption that animals' locations are independent, but in truth animals seldom, if ever, distribute themselves with complete disregard for the locations of other animals. They might tend to be near other animals, or if they are very territorial, they might tend to be far from other animals of the same species. Design-based variance estimation is unaffected by this; model-based variance estimation with an assumption of independent locations is not.

3.2.5 Relevance of design for likelihood-based inference

Most wildlife estimation methods involve assumptions about an observation model. Purely likelihood-based methods necessarily also involve assumptions about a state model, unless the whole survey region was sampled and there is no heterogeneity (in which case all uncertainty comes from the fact that we may not have observed all the animals within the covered region). Once we know the covered region, the design does not enter maximum likelihood estimation (see Section 4.4). This does not imply that we can ignore the design if we are estimating abundance by maximum likelihood. For the maximum likelihood methods of this book, the two main uses of the design are:

(1) *Improving the robustness of the MLE.* Look at Figure 3.8 again. If we had sampled only strip 1, we would really have no way of telling from the observed count whether animal density increases, decreases, or stays constant as we move east. Sampling strip 2 as well would help a bit, but not as much as if we sampled 1 and 6. Because model-based estimates of N depend on the assumptions of the state model, we

would like to be able to check how reasonable the assumptions are once we observe the data. Sampling as many strips as possible, and spreading the units we sample over the range of the survey region, helps us do this.

(2) *Robust variance estimation.* Variance estimates based on assumptions of independent locations of animals will often be biased because animals do not behave independently of one another. For reasons to do with ease of analysis, most spatial state models involve this sort of assumption of independence. An effective compromise strategy is to estimate N by maximum likelihood, but to use design-based methods for variance and confidence interval estimation. We illustrate this strategy in later chapters.

Note that in order to use design-based variance estimates, we need to have selected our sample according to some random design, which is then used to estimate variance. This requirement conflicts slightly with the requirement for robust estimation in (1) above, which would have us spread samples more systematically over the survey region. In practice, both requirements can be met to a large degree provided the number of units sampled is not small, in which case random designs are likely to give good coverage of the survey region, and systematic designs are likely to yield estimates of variance with low bias if they are assumed to be random.

3.3 Summary

This chapter introduced the two fundamental components of the likelihood functions which will be used to estimate abundance and related parameters. They are:

State models: These are statistical representations of the biological, demographic and other processes determining the distribution of animals in space, their sex, their size, etc. – in short, all relevant animal characteristics. We use the vector $\underline{x}$ to refer to these characteristics. Animal abundance, N, is normally an unknown parameter of the state model.

Observation models: These are statistical representations of how animals are detected, given their characteristics. They depend both on the animal characteristics and on the characteristics of the survey effort used to detect the animals (number of observers/traps, identity of observers/type of trap, etc.). We use the vector $\underline{l}$ to refer to the survey characteristics.

Given which parts of the survey region were searched, the state model and observation model together represent our model of how the observed data were generated. Estimation of N is not based on the survey design (unlike the case of purely design-based estimation). However, variance estimates based on the survey design may be more robust than model-based variance estimates obtained by assuming independent locations of animals. Reliable model-based inference requires an ability to check model assumptions, and a large, representative sample makes this easier.

4
Plot sampling

Key idea: "scale up" count from covered area to survey area.

Number of surveys:	Single survey is adequate.
State model:	Animals are assumed to be distributed uniformly and independently in the survey region for model-based likelihood function, but not design-based inference.
Observation model:	All animals in the covered region are detected with certainty.

4.1   Introduction

We refer to quadrat sampling, strip sampling and similar methods under the general name of "plot sampling". Although they may be different in their implementation, they are almost identical from a statistical point of view. The difference between them is in the shape of the plot. Quadrat and strip sampling involves rectangular plots, but circular plots are common

---

Likelihood function and key notation:

$$L(N) = \binom{N}{n} \pi_c^n \, (1 \, - \, \pi_c)^{N-n}$$

N: number of animals in the population (abundance).
n: number of animals counted in the covered region.
$\pi_c = \frac{a}{A}$:   probability that an animal is in the covered region.
a: surface area of covered region.
A: surface area of whole survey region.

---

in some applications. Their essential feature from a statistical perspective is that all animals within the searched plots (whatever their shape) are detected. We use the term "**covered region**" to refer to those parts of the survey region that are searched (i.e. fall inside the sampled plots). We use the term "covered area" to refer to the surface area of the covered region.

When the whole survey region is searched (covered region=survey region) and all animals are detected, this is called a census, or complete count. We don't deal with complete counts in this book – they require no statistical analysis. The plot surveys we deal with are those in which the covered region is smaller than the survey region. That is, a subset of plots or strips within the survey region is chosen according to some design, and only these are searched. The problem is then to estimate the total number of animals in the survey region from the animals you detect in the covered region. As we noted in Chapter 3, there are two quite different approaches to doing this. Design-based methods use the randomness introduced in selecting plots to be searched (i.e. the survey design) as their basis. Model-based methods use a statistical model of the distribution of animals in the survey region as their basis. In keeping with our approach in the rest of this book, we concentrate on model-based methods for plot surveys. Details of design-based plot sampling can be found in any good textbook on sampling theory (Thompson, 1992, for example). Note that the design-based approach is valid whatever the spatial distribution of animals through the survey region.

4.2   A simple plot survey

Suppose we sample from our example population using simple random sampling of strips running the full north–south range of the survey region. The six pre-defined plots or strips from which we sample are shown in Figure 4.1, together with the locations of all animals in the four chosen strips – strips 1, 3, 5 and 6. Our example is rather atypical, but serves to illus-

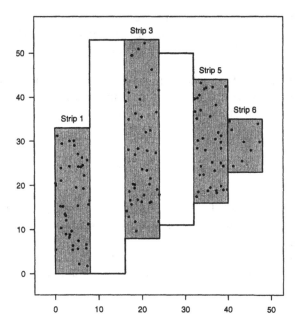

Figure 4.1. Strip survey of example data showing the four covered strips (shaded) and the locations of animals within them.

trate the methods simply. In practice the survey region would usually be divided into many more strips and a much smaller fraction of them would be sampled.

We selected our sample by "simple random sampling" (srs) of plots, as described in Chapter 3.

4.3   Estimation by design

There are many design-based methods for estimating abundance from our survey data. Since this is not the main focus of this book, and the topic is well covered in other books, we consider only the Horvitz–Thompson estimator here. We choose this estimator because (a) it is intuitively easy to understand, (b) you can use it with any survey design, and (c) it can readily be used with an estimate of p when p is unknown.

4.3.1   Horvitz–Thompson estimator of abundance

The probability that any particular plot appears in our srs of size 4 is 4/6. We call the probability that plot j is included in our sample its "inclusion probability", and we write it p_{dj} (the subscript d is for *design*). We call the surface area of the covered region (strips 1, 3, 5 and 6 in our example) the "**covered area**", and denote it a. In our survey, $a = 944\text{m}^2$ and the whole survey area is $A = 1680\text{m}^2$, so the covered area is about 56.2% of the survey area.

If we write the number of animals in the jth covered plot as N_{cj} (the subscript c is for covered), the general form for the Horvitz–Thompson estimator of animal abundance is

$$\hat{N} = \sum_{j=1}^{J} \frac{N_{cj}}{p_{dj}} \tag{4.1}$$

where J plots are sampled. The estimator was proposed by Horvitz and Thompson (1952). Further details and variance estimators can be found in any good sampling theory textbook.

The Horvitz–Thompson estimator of animal abundance with our four-plot survey, in which we observed 42, 44, 38 and 8 animals in strips 1, 3, 5 and 6 respectively, is therefore

$$\hat{N} = \frac{42}{\frac{4}{6}} + \frac{44}{\frac{4}{6}} + \frac{38}{\frac{4}{6}} + \frac{8}{\frac{4}{6}} = 198 \tag{4.2}$$

Table 4.1 shows the estimator for all possible 4-sample srs surveys of these 6 strips, together with the true mean and variance of the estimator. (Note that you can't construct this table unless you know the number of animals in every strip; you would not know the true mean and variance from a survey.)

Points to note:

- The Horvitz–Thompson estimator is design unbiased – its mean over all possible 4-sample srs surveys is exactly equal to the true number of animals in the survey region (250). It can be shown that the estimator is unbiased (Horvitz and Thompson, 1952; Thompson, 1992).

- The MLE (tabulated in the last column) is not design unbiased (i.e. the average over all possible samples given the design is not equal to the true value). However, in this particular case the bias is negligible. Maximum likelihood estimators are not in general unbiased, although they are consistent. That is their bias tends to zero as sample size increases.

- The MLE is identical to the design-based ratio estimator in which the strip surface area is the "auxiliary variable" (see Thompson, 1992, for more on ratio estimators). While the ratio estimator is not unbiased with small samples, its bias tends to zero as sample size tends to

Table 4.1. The sampling distributions of two estimators: $\hat{N}_{HT}$ is the Horvitz–Thompson estimator, and $\hat{N}_{MLE}$ is the maximum likelihood estimator. A is the total stratum area, a is the area of the covered strips. The sample we chose by simple random sampling is in bold.

Sample no.	Strips included	Animals counted	a	$\pi_c = \frac{a}{A}$	$\hat{N}_{HT}$	$\hat{N}_{MLE}$
1	1,2,3,4	204	1360	0.8095	306.0	252.0
2	1,2,3,5	188	1272	0.7571	282.0	248.3
3	1,2,3,6	158	1144	0.6810	237.0	232.0
4	1,2,4,5	198	1224	0.7286	297.0	271.8
5	1,2,4,6	168	1096	0.6524	252.0	257.5
6	1,2,5,6	152	1008	0.6000	228.0	253.3
7	1,3,4,5	178	1160	0.6905	267.0	257.8
8	1,3,4,6	148	1032	0.6143	222.0	240.9
9	**1,3,5,6**	**132**	**944**	**0.5619**	**198.0**	**234.9**
10	1,4,5,6	142	896	0.5333	213.0	266.3
11	2,3,4,5	200	1320	0.7857	300.0	254.5
12	2,3,4,6	170	1192	0.7095	225.0	239.6
13	2,3,5,6	154	1104	0.6571	231.0	234.3
14	2,4,5,6	164	1056	0.6286	246.0	260.9
15	3,4,5,6	144	992	0.5905	216.0	243.9
Mean					250.0	249.9
Std. dev.					34.0	12.1

infinity. If the auxiliary variable (strip area in this case) is highly positively correlated with abundance in the strip (as it is in this case), the variance of the ratio estimator can be considerably smaller than that of the Horvitz–Thompson estimator.

- The higher variance of the Horvitz–Thompson estimator in this example should not be viewed as a reason to prefer model-based estimation. Simple random sampling is not a very good method for selecting strips in this example because strip sizes vary widely. Sampling with probability proportional to size, for example, will improve the performance of the design-based estimator.

4.4 Maximum likelihood estimation

A design-based approach treats the positions of animals in the survey region as fixed. If you selected sample 9 in our example, you would see 132 animals – there is no randomness involved at this stage. The only randomness is that introduced by the design (i.e. which strips you choose). This is of no use at all with maximum likelihood estimation, for the following reason.

Remember that the likelihood function tells us how likely any value of N is, given the data we observed. But when we treat the number and positions of animals as fixed, the probability of getting the data we did depends only on the design, and the design does not depend on N. So we are equally likely to get the data we did, for all possible values of $N \geq n$; likelihood functions constructed on the basis of conventional survey designs contain no useful information about N! There are designs for which this is not the case; Thompson and Seber (1996) discuss this sort of design.

The likelihood-based approach is founded on the randomness in the locations of the animals, and this is specified by the state model. (There is no randomness in the observation model for plot sampling because detection is certain within the covered region.) Conceptually, animals' positions are governed by the true (unknown) state model, and their positions at the time of our survey are just a realization of this random process.

4.4.1   Point estimation

If a animal is equally likely to fall anywhere in the survey region (with area A), it follows that the probability that it falls in a covered region with area a is just $\pi_c = a/A$. If animals' locations are independent and we think of the event "animal falls in the covered region" as a "success", then n, the number of animals in the covered region, is a binomial random variable with parameters N and π_c:

$$L(N) \;\; = \;\; \binom{N}{n} \pi_c^n \, (1 \, - \, \pi_c)^{N-n} \tag{4.3}$$

The MLE for N is just $\hat{N} = n/\pi_c = nA/a$. So the likelihood for N, the number in the survey region, given that $n = 132$, is Equation (4.3) with $\pi_c = a/A = 944/1680 = 0.5619$. The MLE is $\hat{N} = \frac{132 \times 1680}{944} = 235$, to the nearest integer. Using the variance estimator obtained in Section 2.2.4, we can estimate the standard error of $\hat{N}$ as

$$\hat{se}\left[\hat{N}\right] \;\; = \;\; \sqrt{\frac{235 \times (1 - 0.5619)}{0.5619}} \approx 13.5 \tag{4.4}$$

4.4.2   Interval estimation

Parametric estimation

If $\hat{N}$ was assumed to be normally distributed, our estimated 95% confidence interval, $\hat{N} \pm 1.96\hat{se}[\hat{N}]$, would be (208, 261), to the nearest integer.

We can estimate a 95% confidence interval without the assumption of normality using the assumed binomial distribution of n to get exact, profile likelihood, or parametric bootstrap CIs. These are all based on the state model of Equation (4.3) with $\pi_c = 944/1680$, and in the case of the parametric bootstrap, $\hat{N} = 235$. An estimated confidence interval from 9999 bootstrap resamples is (213, 264).

The normal approximation works quite well and there is not much difference between the bootstrap and normal CIs. The profile likelihood CI is (209, 262). In this case all methods give CIs that are quite close to the exact CI, which is (211, 264).

Because we know that $N = 250$ in this example, we can evaluate the true variance and standard deviation of $\hat{N}$ under the assumptions of the state model. The standard deviation is $\sqrt{250 \times (1 - 0.5619)/0.5619} \approx 14$ animals. The reason that this is so close to the standard deviation of the MLEs of Table 4.1 is that in this particular case, the assumption that animals locate themselves independently between strips holds. If they were aggregated, or there was trend in density across strips, the variance estimate obtained under the assumption of independent animal locations would on average be too low, and the estimated confidence intervals would be too narrow. We deal with an example in which this is the case below.

Nonparametric bootstrap

If we apply the nonparametric bootstrap by resampling the units of the design (the strips here), we need not assume independent animal location – which was implicit in the parametric bootstrap above. In practice, we have too few strips in our example to make this feasible, but in more realistic applications, with good design, this should not be the case. We chose to have few strips in our example in order to keep it simple for illustration.

A nonparametric bootstrap could be applied in a number of ways. We could, for example, ensure that each bootstrap replicate samples approximately the same proportion of the survey region as the original sample. Alternatively, we could ensure that replicates all have the same number of strips as the original sample. The latter is simpler and involves the following steps. (When plot sizes are all equal, the two methods above are identical.)

(1) For replicate i ($i = 1, \ldots, B$), choose J plots from those sampled, with replacement, where J is the number of sampled plots in the original design.

(2) Calculate the MLE for replicate i: $\hat{N}_i = n_i/\pi_{c_i}$, where n_i is the total number of animals in the J resampled strips, a_i is the sum of the areas of the resampled strips, and $\pi_{c_i} = a_i/A$. Store $\hat{N}_i$.

(3) Steps (1) and (2) are repeated B times, where typically, B might be between 400 and 2,000.

(4) Calculate the mean and variance of the B $\hat{N}_i$s, as follows:

$$\hat{N}_b = \frac{\sum_{i=1}^{B} \hat{N}_i}{B} \tag{4.5}$$

$$\widehat{Var}_b\left[\hat{N}_b\right] = \frac{\sum_{i=1}^{B}\left(\hat{N}_i - \hat{N}_b\right)^2}{(B-1)} \tag{4.6}$$

(5) Estimate a 95% CI for N by $(\hat{N}_{(j)}; \hat{N}_{(k)})$, where $j = (B+1) \times 0.025$, $k = (B+1) \times 0.975$ and the notation $\hat{N}_{(j)}$ and $\hat{N}_{(k)}$ indicate the jth and kth largest $\hat{N}_i$s, respectively.

The point estimate based on the original data, $\hat{N}$, is usually used in preference to the bootstrap mean, $\hat{N}_b$. $Var[\hat{N}]$ is estimated by $\widehat{Var}_b[\hat{N}_b]$.

If our design involved sampling strips without replacement, this bootstrap method will give a positively biased estimate of variance because it does not take account of the finite population correction factor (fpc). The fpc adjusts for the fact that when sampling is without replacement, uncertainty shrinks quicker as sample size increases than when sampling is with replacement, because every new plot sampled adds information about animals not yet encountered. For example, suppose that the survey region was divided into T plots. If we chose a sample of T plots with replacement, our sample would contain fewer than T different plots on average, because some plots would be chosen more than once in a single sample; there would be uncertainty about the number of animals in plots that were not chosen. If we chose a sample of T plots without replacement, we would be certain of seeing the whole population in any single sample, leaving no uncertainty at all about the population size. In this sense, sampling without replacement gives more information per sample unit than does sampling with replacement. It is possible to get variance and confidence interval estimates using bootstrap methods which do incorporate an appropriate adjustment for sampling without replacement. See Davison and Hinkley (1997, pp 92–100) for details.

We note in passing that since the MLE is identical to the ratio estimator with strip area as an auxiliary variable, design-based variance estimators are available from finite population sampling theory. For example, if sampling is with replacement:

$$\widehat{Var}_{wr}\left[\hat{N}\right] = T^2 \frac{\sum_{j=1}^{J}\left(n_i - \frac{n}{a}a_i\right)^2}{J(J-1)} \tag{4.7}$$

where n_i is the number of animals in the ith strip, n is the number in all sampled strips, a_i is the area of the ith strip, and the subscript wr is for "*with* replacement". If sampling is without replacement then

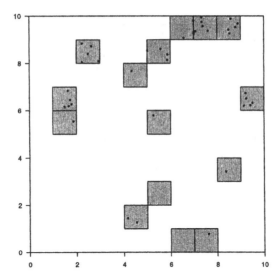

Figure 4.2. A plot survey of a simulated animal population. Shaded rectangles were sampled. Dots represent detected animals.

Table 4.2. Sampled plots and associated counts. Sampled plots are indexed by the x-coordinate and y-coordinte of their bottom left corner.

x-coord.	6	7	8	2	5	4	1	9	1	5	8	5	4	6	7
y-coord.	9	9	9	8	8	7	6	6	5	5	3	2	1	0	0
count (n)	2	7	5	5	4	1	5	5	1	1	1	0	2	0	1

$$\widehat{Var}_{wor}\left[\hat{N}\right] \;=\; \left(1 - \frac{J}{T}\right)\widehat{Var}_{wr}\left[\hat{N}\right] \tag{4.8}$$

4.4.3 A more realistic example

We use the R library WiSP to simulate a population in a 10 km × 10 km survey region, and a plot sample survey of it. Fifteen plots were chosen with replacement and the number of animals in each counted. The plots and detected animals are shown in Figure 4.2. The counts are summarized in Table 4.2.

A total of 40 animals was detected and the MLE is $\hat{N} = 267$ to the nearest integer. A parametric bootstrap with binomial parameters $\hat{N} = 267$ and $\pi_c = 15/100$, and 999 replicates, gives a percentile method 95% confidence interval for N of (193; 340). A nonparametric bootstrap, based on resampling the 15 selected grid squares, gives a percentile method 95% confidence interval for N of (160; 380). Equation (4.7) gives a design-based

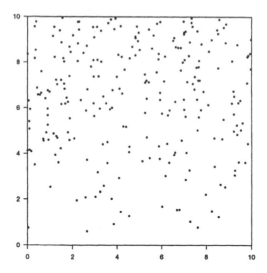

Figure 4.3. A simulated animal population of 250 animals, with north–south trend in density.

95% confidence interval for N (assuming $\hat{N}$ is normally distributed) of (153; 381). The bootstrap methods have the advantage that they do not rely on the assumption of normality for estimating the confidence interval.

The exact and profile likelihood confidence intervals are (202, 356) and (198, 351), respectively. Note that the widths of the parametric bootstrap, exact and profile likelihood CIs are very similar (147, 154 and 153), while the design-based and the nonparametric bootstrap CIs are about 25% wider (228 and 220 respectively).

All but the design-based and nonparametric bootstrap CIs assume that animals are located independently in the survey region; they assume a uniform spatial state model. These results suggest that this assumption is not valid. Further evidence that this is the case is provided by looking at Figure 4.2 again. The plots in the north clearly contain a higher number of animals on average than do those in the south — suggesting a north–south trend in density, not a uniform density throughout the region.

With this evidence, it would be unwise for variance estimation to rely entirely on a model which assumes uniform density throughout the survey region.

The simulated population does have a strong north–south trend in density, as is apparent from the plot of the locations of all its members in Figure 4.3 – clearly a uniform state model is not appropriate.

Note that the maximum likelihood estimate is quite close to the true N (which is 250): with a random survey design, the point estimate does not depend heavily on the assumed uniform state model. By contrast, the

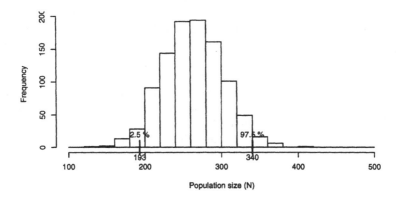

Figure 4.4. Simulated data: the histogram is the distribution of $\hat{N}$, estimated by parametric bootstrap with $\hat{N} = 267$ and $\pi_a = 0.15$. The short vertical lines indicate the 95% CI bounds for the percentile method.

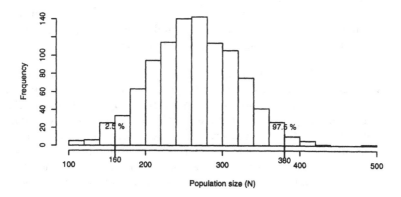

Figure 4.5. Simulated data: the distribution of $\hat{N}$, estimated by nonparametric bootstrap, resampling plots. The short vertical lines indicate the 95% CI bounds for the percentile method.

parametric estimate of variance will be negatively biased when animals occur at higher density in some parts of the survey region, because of its sensitivity to the assumed independence of locations of animals. This is also the reason that the parametric bootstrap, exact and profile likelihood CIs are so much narrower than the nonparametric bootstrap CI. The difference in the distribution of $\hat{N}$, with and without the assumption that animals are located independently, can be seen by comparing Figure 4.4 and Figure 4.5. The nonparametric bootstrap incorporates the additional variance resulting from non-random animal locations, as does the design-based method; the other methods do not. As a result, the nonparametric and design-based confidence intervals are more reliable here.

Note that we could use the parametric bootstrap with a model which does not involve the assumption that animals are distributed uniformly and independently of one another.

4.5   Effect of violating assumptions

Assumption 1 All animals in the covered region are detected.

Effect of violation: Estimator is negatively biased by a factor equal to the probability that an animal in the covered region is detected.

Assumption 2 Animals are distributed uniformly and independently[1] of one another. Most animal populations do not distribute themselves independently of one another; they tend to cluster or, in the case of highly territorial animals, to avoid one another.

Effect of violation: CIs and variances based on the assumption that animals are uniformly distributed will be too narrow (if animals cluster), or too wide (if animals are more regularly distributed). They will be similarly biased if the assumption of independent location of animals is violated: too narrow if there is positive correlation in their locations, too wide if there is negative correlation. The nonparametric boostrap method with plots as the resampling unit is not based on these assumptions and is robust (insensitive) to their violation.

---

[1]When we say that animals distribute themselves uniformly, we mean that they are equally likely to be anywhere in the survey region, not that the distances between animals are all the same. If they are uniformly and independently distributed, then any one animal is equally likely to be anywhere in the survey region, and the location of one animal has no effect on the location of other animals.

4.6 Summary

Plot surveys are unusual among wildlife abundance estimation methods in that they **do not involve an observation model**. The basic assumption of the method is that **all animals in the covered region are detected**.

All uncertainty in estimating abundance therefore arises from the fact that the whole survey area is not searched. There are two approaches to dealing with this uncertainty. A design-based approach allocates all randomness to the survey design, which is in the control of those designing the survey. A model-based (likelihood-based) approach allocates all uncertainty to a state model which is assumed to determine the distribution of animals in the survey region. The major advantage of the former approach is that it involves no assumptions about the unknown processes determining the distribution of animals. The major advantage of the latter approach is that it can improve precision and it allows animal density to be estimated as a smooth function of position (or other covariates). Like any model-based approach, estimates can be misleading if assumptions are invalid, so it is important that they be verified as far as possible. With a sufficiently large sample size, and random selection of plots (or systematic selection with a random start point), this is possible.

Only uniform density state models were considered in this chapter; the real benefits of using a state model are likely to come when animal distribution is not uniform. This sort of model is covered in Part III.

4.7 Exercises

Exercise 4.1   Figures 4.6 and 4.7 show the results of two surveys of populations in two 100 m × 100 m survey regions. Summary statistics from each survey are shown in Table 4.3.

(a) Estimate the animal abundance in each survey region.

(b) Estimate a 95% confidence intervals for animal abundance under the assumption that animals are uniformly and independently distributed in each survey region.

(c) [2]Estimate a 95% confidence intervals for animal abundance using the percentile method and a nonparametric bootstrap with plots as sampling units.

---

[2]You can use the R library WISP to answer this question (the surveys are on the website given at the end of Chapter 1).

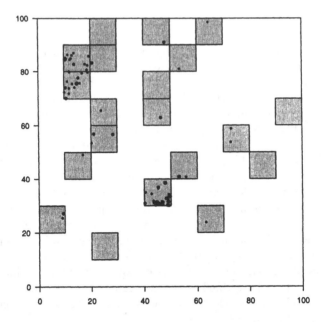

Figure 4.6. Results of plot survey 1. Shaded squares are searched plots; dots are detected animals.

Table 4.3. Summary statistics from plot surveys 1 and 2.

Statistic	Survey 1	Survey 2
Number of plots sampled	20	20
Individual plot size	$100m^2$	$100m^2$
Number of animals detected	70	51
Mean number of animals per plot	3.5	2.55
Survey area	$10,000m^2$	$10,000m^2$
Proportion of survey area covered	20%	20%

(d) For each survey, explain the difference between the confidence intervals from (b) and (c), and say which is the more reliable method.

Exercise 4.2   Table 4.4 shows the plot sizes and the number of animals in each plot, for a plot survey of an animal population. The six plots cover the whole survey region. To estimate the population size, four plots are selected without replacement.

(a) Construct a table similar to Table 4.1 for all possible outcomes of a survey of this population consisting of four of the plots, chosen randomly with equal probability and without replacement.

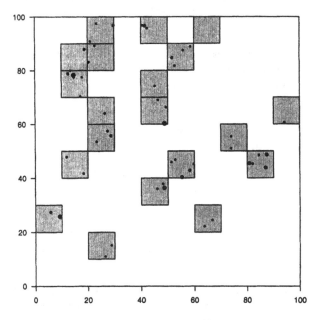

Figure 4.7. Results of plot survey 2. Shaded squares are searched plots; dots are detected animals.

Table 4.4. Six plots spanning the survey region, and the number of animals in each plot.

Plot number	1	2	3	4	5	6
Plot area	50	100	100	200	200	350
Count (n)	62	42	35	30	20	11

(b) Hence show that the Horvitz–Thompson estimator of N is design-unbiased, but the MLE is positively biased by about 7%.

(c) Show also that the standard error of the MLE is much larger than that of the Horvitz–Thompson estimator in this case.

(d) If we assume that animals locate themselves independently of one another and each animal is equally likely to be anywhere in the survey region, what is the variance of the MLE of N? Is this equal to, less than, or more than the true variance for this population? Why?

Exercise 4.3 Figure 4.8 shows the sampling units for a plot survey of a simulated plant population.

(a) By choosing four of the plots randomly with equal probability and without replacement, survey the population and estimate abundance and a 95% confidence interval for N.

(b) Prove that if the plots are all the same size with the above sampling scheme, then the MLE and Horvitz–Thompson estimators are identical.

(c) If $N = 200$ and the likelihood Equation (4.3) applies (and we take the covered region as given), what is the variance of the MLE? Is this equal to, less than, or more than the true variance of the MLE for this population? Why?

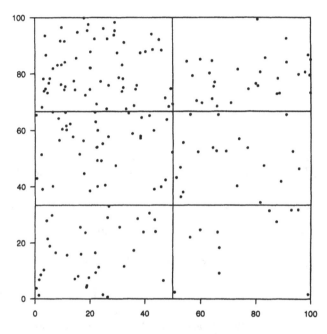

Figure 4.8. A plant population and the plot sampling frame comprising six plots, which is to be used to estimate abundance by plot sampling.

5
Removal, catch-effort and change-in-ratio

Key idea: many captures after removals implies a population much larger than the number removed, and vice versa.

Number of surveys:	At least two surveys required.
State model:	None. Assume complete coverage.
Observation model:	Detection (capture) probability is the same for all animals on all occasions, or depends only on survey–level variables (measures of survey effort). Independent detections (captures).

5.1 Introduction

In Chapter 2, we introduced maximum likelihood estimation for the case where abundance (N) was unknown, but detection probability (p) was known (or at least assumed). In Chapter 4, we dealt with the only wildlife survey methods in which p is known – and is equal to 1. In this chapter we consider for the first time the more common scenario in which neither N

Likelihood function and key notation:

$$L(\underline{N}_1, \underline{\theta}) = \prod_{s=1}^{S} \prod_{\underline{x}} \binom{N_s(\underline{x})}{n_s(\underline{x})} p(l_s)^{n_s(\underline{x})} \left(1 - p(l_s)\right)^{N_s(\underline{x}) - n_s(\underline{x})}$$

$\underline{x}$:   vector of animal-level variables (animal "type").

$\underline{N}_1$:   vector of initial abundances of each animal "type" (see Equation (B.11)).

$N_s(\underline{x})$:   abundance of animals of type $\underline{x}$ on survey s.

$n_s(\underline{x})$:   number of animals of type $\underline{x}$ detected (captured) on survey s.

l_s: survey effort type on survey s.

$p(l_s)$:   probability of detecting (catching) an animal on survey s, using effort type l_s.

$\underline{\theta}$:   parameters of the detection (capture) function, $p(l_s)$.

nor p are known. We again use the data from our example population to illustrate the problem and the methods.

The data we get from the first observer's survey are shown in Figure 2.14 — a total of 64 "captures".[1] Note that the observer searched the whole survey region, so we do not need a state model to make inferences about numbers in any uncovered region, and the design does not introduce any randomness. (The design is "search everywhere".) In other words, all the randomness is contained in the observation model.

We start by using the same observation model as we used in Chapter 2, but with unknown p:

(1) All animals are equally catchable, and are caught with probability p.

(2) Animals are caught independently of one another.

The only useful bit of data we have from the first survey is $n = 64$. But we have no way of knowing whether n is a large proportion of a low abundance (large p, small N), or a small proportion of a large abundance (small p, large N). We can't estimate two unknown parameters (N and p) with only one bit of data (n).

Our assumptions above lead to a binomial likelihood function, as they did before. The difference is that it is now a function of two parameters (N and p). We would like to use the maximum likelihood method to find the

---

[1] With a brief exception at the start of Section 5.3, we assume throughout this chapter that all detected animals are captured, and use the terms "capture" and "detection" interchangeably.

unique N and p which are most likely, given that $n = 64$. But to estimate both N and p, we need additional data and/or additional assumptions. In Chapters 2 and 4, we made strong assumptions about p. In this chapter we start to look at methods which use data from additional surveys, together with weaker assumptions about p. We deal with the simple removal method first.

Note that throughout this book we assume that the number of removals is known with certainty. In practice this is not always the case; there may be bias in reporting catch (if part or all of the catch was illegal, for example), and there may be uncertainty about catch (because only a sample of the catch is observed, for example).

5.2   Removal method

5.2.1   Point estimation

The method involves removing a known number of animals on, or after, the first survey. The population is then surveyed a second time, keeping detection probability constant. The size of the reduction in the number of detections between first and second survey occasions contains information about N (and detection probability). For example, if the number of detections remains about the same on the two occasions, despite having removed a large number of animals between survey occasions, this suggests that the number of animals removed was a negligible proportion of the number there. Hence the number there must have been large.

We use our example population for illustration. Suppose observer 1 removed the 64 animals she detected. Of the remaining animals, observer 2 detected 33. We now have two bits of data: $n_1 = R_2 = 64$, and $n_2 = 33$. (Recall from Chapter 3 that R_s is used to denote the total number of removals by the start of survey occasion s.)

What does the new piece of information, $n_2 = 33$, tell us? If observers 1 and 2 saw exactly the same proportion of the population present at the time of their surveys, then the number of animals present at the time of the second survey (which we refer to as N_2) must have been 33/64ths of the number there at the time of the first survey (which we refer to as N_1 or just N). That is

$$N_2 \;=\; \frac{33}{64} N \tag{5.1}$$

We also know that this number is 64 less than the number of animals there to start with. That is

$$\frac{33}{64} N \;=\; N - 64 \tag{5.2}$$

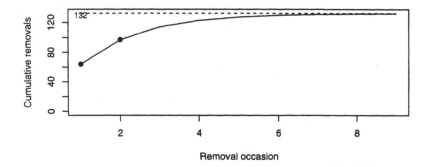

Figure 5.1. Observed (dots) and predicted (solid line) cumulative removals from a two-sample removal survey of the St Andrews example population. The height of the dashed line is the estimated abundance.

Solving this equation for N gives us the estimate $\hat{N} = 64^2/(64-33) = 132$ (to the nearest integer). In general, for any n_1 and n_2, we get the estimator

$$\hat{N} = \frac{n_1^2}{n_1 - n_2} \tag{5.3}$$

Figure 5.1 shows graphically how the estimator works. The constant removal probability, p, and abundance, N, together specify a curve of predicted cumulative removals of the shape shown in the figure. The estimator finds the curve (i.e. the N and p) that fits the observed removals best in terms of likelihood. The point at which the curve flattens off is the point at which the model predicts that the whole population will have been removed (more removal occasions lead to no more removals). The height of the curve at this point is the estimated abundance.

The true size of this population is $N = 250$. So our estimate of 132 is far off target, but is this bias or just random variation? We return to this issue below. First we summarize how we arrived at the estimator. We

(1) gathered an additional bit of data n_2, and

(2) assumed that both observers saw exactly the same proportion of the animals present.

While (2) does not assume that we know the proportion, it is still an implausibly strong assumption. A slightly more plausible assumption is that the two observers are equally good, *on average*. That is, we assume that both observers have the same p, although on any particular survey, observer 1 might detect a higher proportion than observer 2, or vice versa. This assumption is the basis of the simple two-sample removal method. It leads to a maximum likelihood estimator which is identical to the estimator we derived above. In general, there may be more than two sampling occasions, in which case deriving estimators by intuition is rather more difficult and the formal maximum likelihood method comes into its own.

5.2.2   Simple removal method MLE

Technical
section
↓

For a given survey occasion, we assume (1) all animals are detected (captured) with the same probability p, and (2) detections are independent events. The likelihood for N and p from a single survey occasion is thus given by Equation (2.3), with both N and p now unknown. With the additional assumptions that

(3) detections (captures) are independent between occasions, and

(4) detection (capture) probability, p, is the same on all occasions,

the likelihood for N and p is just the product of the likelihoods for the individual occasions (taking due account of removals). With S survey occasions in all, in which a total R_s animals were removed by the start of occasion s, and n_s animals were detected on occasion s, the likelihood is

$$L(N,p\,|\,\{n,R\}) \quad \propto \quad \prod_{s=1}^{S} L(N,p\,|\,n_s,R_s)$$

$$= \quad \prod_{s=1}^{S} \binom{N_s}{n_s} p^{n_s} (1-p)^{N_s-n_s} \quad (5.4)$$

where $\{n,R\} = (n_1,R_1),\dots,(n_S,R_S)^2$, $N \equiv N_1$ (we omit the subscript for brevity) and $R_1 = 0$.

From this we get the MLEs, which satisfy the following two equations:

$$1 - \frac{n}{\hat{N}} = (1-\hat{p})^S \quad \text{and} \quad \hat{p} = \frac{n}{\sum_{s=1}^{S}(\hat{N}-R_s)} \quad (5.5)$$

where n is the total number of detections over all S occasions (see Exercise 5.1). It is then easy to show that the MLE of N for

---

[2]Notation: we use vectors to refer to animal abundance by type and braces to refer to animal abundance by occasion. So $\{n,R\}$ denotes sets of counts and removals by occasion (with only one animal type), while (n_s, R_s) denotes a set of counts and removals of each type (on the single occasion s)

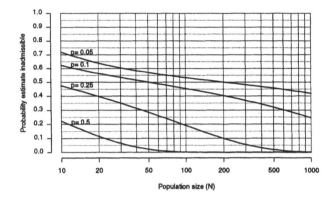

Figure 5.2. The probability that the two-sample removal method estimator is inadmissible (because $\hat{p} \leq 0$).

the two-sample simple removal method is identical to the intuitive estimator of Equation (5.3), and that the corresponding MLE for p is

$$\hat{p} \;=\; \frac{n_1 - n_2}{n_1} \tag{5.6}$$

(see Exercise 5.2).

↑
Technical
section

It is apparent that the two-sample removal method estimator is inadmissible whenever $n_2 \geq n_1$ because the probability p cannot be negative; nor can it be zero (which happens when $n_2 = n_1$). This failure can occur even if all the model assumptions are met and thus constitutes a limitation of the method. The method is unlikely to yield good estimates when the detection probability p is small. Indeed there is a substantial probability that $n_2 \geq n_1$ when p is small. This is illustrated in Figure 5.2, which gives the probability of obtaining inadmissible estimates for a range of population sizes N and detection probabilities p. Thus for example, for a population of size $N = 200$ and a detection probability of $p = 0.1$, an inadmissible estimate will be obtained about 25% of the time.

5.2.3   Interval estimation

All uncertainty arises from the observation process, and variance and confidence interval estimates are based on the observation model alone. This is the product of binomial pdfs (Equation (5.4)). Seber (1982, p 312) gives the following expressions for the asymptotic variances of $\hat{N}$ and $\hat{p}$

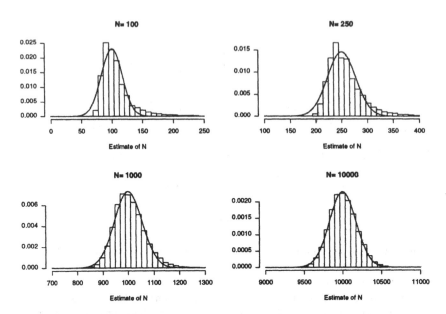

Figure 5.3. Distribution of the simple removal method estimator $\hat{N}$ for $p = 0.5$ and various values of N. The histogram is the distribution from 20,000 simulations; the curve is the normal approximation.

$$Var[\hat{N}] \quad = \quad \frac{N(1 - q^S)q^S}{(1 - q^S)^2 - (pS)^2 q^{S-1}} \tag{5.7}$$

$$Var[\hat{p}] \quad = \quad \frac{(qp)^2(1 - q^S)}{N\left[q(1 - q^S)^2 - (pS)^2 q^S\right]} \tag{5.8}$$

where $q = (1 - p)$.

Variances could be estimated using these equations with $\hat{N}$ and $\hat{p}$ replacing N and p, and confidence intervals could then be estimated assuming normality of $\hat{N}$ and $\hat{p}$. In the survey of our example population, this results in a 95% confidence interval for N of $(90; 175)$. This interval is worrying in two respects. First, the lower interval is below the total number of animals detected (97). Second, we know that the true number of animals is 250. We expect that a 95% CI should include the true N 95% of the time. Is this just one of the 5% of occasions in which it does not, is the estimator biased, or does the normal approximation CI have a lower coverage probability than 95%? (The "coverage probability" of a CI is the probability that it includes the thing being estimated.)

We investigate this question using simulation. The CI estimate is based on the theoretical result that the distribution of $\hat{N}$ approaches the normal as N approaches infinity. In a real population, N is necessarily finite, and

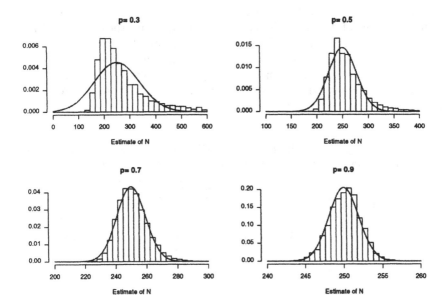

Figure 5.4. Distribution of the simple removal method estimator $\hat{N}$ for $N = 250$ and various values of p. The histogram is the distribution from 20,000 simulations; the curve is the normal approximation.

we have no guarantee that $\hat{N}$ is indeed normally distributed, nor that the (asymptotic) variance $\hat{N}$, given in Equation (5.7), is accurate. One simple way to assess these issues is to compute $\hat{N}$ and the confidence interval for N for each of a large number of samples from a population with known N and p.

Figure 5.3 shows histograms of the estimates $\hat{N}$ that were obtained by generating 20,000 samples for $p = 0.5$ and for $N = 100$, 250, 1,000 and 10,000. Also shown are the corresponding normal distributions based on the asymptotic result. We see that the approximation improves as N increases but that it is somewhat inaccurate for $N = 100$ and $N = 250$. The inaccuracy in the tails of the distribution is noticeable and relevant because it is the tails that determine confidence limits.

Figure 5.4 investigates the approximation for $N = 250$ but for different values of p, namely 0.3, 0.5, 0.7, 0.9. The figure reveals that the approximation is quite accurate for large p but is very poor for small p. The distribution of $\hat{N}$ is much narrower for large p than it is for small p. (Note the different range of $\hat{N}$-values in the four histograms.) Thus the quality of the approximation depends not only on the population size N but also on the detection probability p.

Table 5.1 gives values of the coverage probabilities obtained in the simulation for a range of N and p. Most noticeable here is that these probabilities differ substantially from 0.95 for small detection probabilities p.

Profile likelihood and percentile method confidence intervals usually perform better than those based on the assumption of normality, so if the problem lies entirely with the normal approximation CI, the profile likelihood or percentile method CIs should do better.

Table 5.1. Coverage probabilities of nominal 95% confidence intervals for N based on asymptotic normality of the simple removal method estimator $\hat{N}$.

	$p = 0.1$	$p = 0.3$	$p = 0.5$	$p = 0.7$	$p = 0.9$
$N = 100$	0.25	0.73	0.90	0.92	0.91
$N = 250$	0.36	0.85	0.93	0.94	0.94
$N = 1,000$	0.55	0.91	0.94	0.94	0.94
$N = 10,000$	0.88	0.95	0.95	0.95	0.94

Profile likelihood confidence interval

So far we have covered profile likelihood confidence intervals with one unknown parameter. In that case the profile likelihood is the same as the likelihood. The simple removal method likelihood involves two parameters: N and the nuisance parameter p.

The profile likelihood for N is the likelihood function evaluated at N and at a particular value of p that can be computed from N. The particular value of p for a given N is denoted by $\hat{p}(N)$. It is the value that maximizes the likelihood for that N. For the simple removal method with $S = 2$ occasions, it is given by

$$\hat{p}(N) = \frac{n_1 + n_2}{2N - n_1} \tag{5.9}$$

(see Exercise 5.4).

Thus for example, given the data and assuming that $N = 140$, the maximum likelihood estimator of p is $\hat{p}(140) = 0.449$.

Figure 5.5 shows the difference between the profile log-likelihood and its maximum. The profile likelihood confidence interval is (106, 221). The lower bound now exceeds the number of animals removed, and the upper bound of this interval is closer to the true N of 250 than was the normal approximation CI, although neither interval covers the true value for these data.

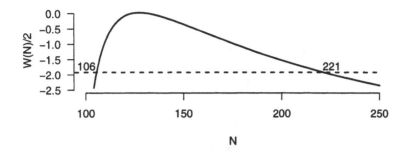

Figure 5.5. $W(N)/2 = l(\hat{N}, \hat{p}) - l(N, \hat{p}(N))$: the difference between the removal method profile log-likelihood function and its maximum, for $115 \leq N \leq 155$, when $n_1 = 64$ and $n_2 = 33$.

Percentile method confidence interval

To implement a bootstrap method of estimating a confidence interval for N, you need to think about the sources of variance. The survey design is "cover the whole survey region" so there is only one designed sampling unit (the whole region), and a simple nonparametric bootstrap using this as the resampling unit is no use because it will reproduce the original data every time. The only source of variance is due to the randomness in the observation process. One way to capture this in a bootstrap procedure is to resample parametrically using our fitted model of the observation process. The idea behind the method is as follows. If we knew N and p, then we could simulate a large number of surveys on the computer (by generating many pairs of values n_1 and n_2) and then compute the corresponding estimates of N. However, we do not know N and p. The parametric bootstrap estimate of a confidence interval for N applies the above procedure, but using $\hat{N}$ instead of N, and $\hat{p}$ instead of p. The bootstrap estimates of N are ordered, and percentile confidence limits obtained. For example, if we generate 999 such estimates, the 25th and 975th largest estimates provide an approximate 95% CI for N.

In our example, $\hat{N} = 132$ and $\hat{p} = 0.484$. To generate the data for a survey, we first generate a realization (for the first occasion), n_1^*, from a binomial distribution with parameters 132 and 0.484, and then a realization (for the second occasion), n_2^*, from a binomial distribution with parameters $132 - n_1^*$ and 0.484. The corresponding estimate for the population size, $\hat{N}^*$, is computed from Equation (5.3).

Suppose that B samples are generated. Denote the ith largest estimate by $\hat{N}_{(i)}^*$. Then the bootstrap estimate of the 95% CI for N is given by $(\hat{N}_{(j)}^*, \hat{N}_{(k)}^*)$, where $j = (B+1) \times 0.025$ and $k = (B+1) \times 0.975$.

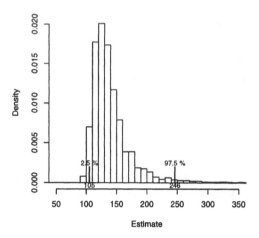

Figure 5.6. Parametric bootstrap distribution of the simple removal method estimator $\hat{N}$; the 95% "percentile" confidence limits for N is indicated by the dashed lines.

Figure 5.6 shows a histogram of $B = 10,000$ parametric bootstrap estimates as well as 250th and 9,751st largest estimates obtained, namely 105 and 246. So the parametric bootstrap 95% CI for N is (105, 246).[3]

The nonparametric bootstrap procedure is very similar in that we also compute B estimates using generated surveys from a population of size $\hat{N}$. The only change is that a different method is used to generate the surveys. Firstly each of the $\hat{N}$ animals is assigned a "capture history" as follows. For the two-sample survey, $\hat{N} - n_1 - n_2$ animals were never captured and are assigned the capture history (0,0); the n_1 captured on the first occasion are assigned (1,0) and the n_2 captured on the second occasion are assigned the history (0,1). To generate a bootstrap survey, we take a simple random sample (with replacement) from the $\hat{N}$ animals and note their capture histories. From these we can compute n_1^* and n_2^* for the survey and hence $\hat{N}^*$. For a two-sample removal survey, the nonparametric bootstrap has no advantage over the parametric bootstrap, but with more occasions, it is better able to accommodate deviation from the assumption that detection probabilities remain constant. In this case, it is more robust than the parametric bootstrap.

The histogram in Figure 5.6 is clearly asymmetric and very different from a normal distribution. Thus the confidence interval based on the normal approximation, which is the only one of the four methods described that will always result in a CI that is symmetric about the estimate $\hat{N}$, will per-

---

[3]Some of the bootstrap surveys yielded $n_2^* \geq n_1^*$ for which $\hat{N}^*$ is inadmissible; these have not been discarded. Providing fewer than 2.5% of estimates are inadmissible, a valid confidence interval can be obtained without discarding them.

form poorly here. The other methods are able to accommodate skewness in the distribution of the estimator and, in this example, give very similar intervals. While not necessarily being symmetric, the profile likelihood and parametric bootstrap contain a strong assumption about the model – namely that detection probability is the same on all occasions. When this assumption fails, as it has in this case (see below), these methods will perform less well. In general (and with more than two occasions), the nonparametric bootstrap contains fewer assumptions and if the assumptions of the model are violated, can perform better than the profile likelihood or parametric bootstrap methods.

In our example, the confidence interval did not cover the true value of N, which was known to be 250. A possible explanation is that the sample is an unusual one. The histograms in Figure 5.4 suggest that the estimate $\hat{N} = 132$ is atypically small for $N = 250$. This may be chance, but an alternative and more plausible explanation is discussed in the sections that follow – namely that one or more assumptions that we made are invalid.

5.2.4 Heterogeneity

There is reason to suspect that the estimator is negatively biased with this example population because the assumption that all animals are equally detectable (i.e. that p is the same for all animals on all occasions) is not reasonable. This population is quite heterogeneous; animals vary in size, distance from the observer, colour and exposure above the grass (see Figure 3.6) and this probably results in some animals being more detectable than others.

The problem is quite general and we illustrate it using the simulated population of size $N = 250$ shown in Figure 5.7, which was created with the library WiSP. Animal exposure (x) in this population has a beta distribution, with highest densities in the vicinity of $x = 0$ (least exposed animals) and $x = 1$ (most exposed animals).

Two two-sample removal surveys of the population were simulated, both with expected sample size $n = n_1 + n_2$ of about 90. In the first, the least exposed animals were detected with probability $\frac{1}{100}$ while the most exposed animals were 40 times more likely to be detected. In the second, all animals were equally catchable.

The first survey, with heterogeneous detection probabilities, generated sample sizes of $n_1 = 61$, $n_2 = 32$. Using Equation (5.3) with these data, we estimate N to be 128 to the nearest integer, with nonparametric bootstrap CI (98; 216).

The survey with homogeneous catchability generated sample sizes $n_1 = 52$ and $n_2 = 39$. Using Equation (5.3) again, $\hat{N} = 208$, to the nearest integer, with nonparametric bootstrap CI (111; 965). Although the confidence interval is very wide, we seem to have done better when there was no heterogeneity. The reason for this can be seen in Figure 5.8, which compares

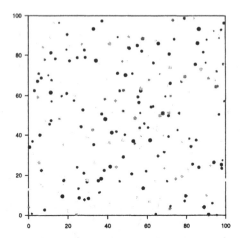

Figure 5.7. A simulated population of $N = 250$ animals. Dots are animals; the darker the dot, the more exposed the animal.

the animal exposures of detected animals in each survey to that for the whole population.

When the more exposed animals are more likely to be detected (the heterogeneous case), the distribution of exposure among sampled animals is notably skewed towards the right compared with the distribution of exposure in the population; animals with high exposure are caught much more frequently than animals with low exposure. As a result, the most catchable animals tend to be removed first, so that the average catchability of uncaught animals is lower after each removal. Because the likelihood and the estimator incorporate an assumption of constant detection probability, the likelihood function must be maximized by reducing N until it is small enough to account for the reduced sample size on the second occasion (which is really due to reduced catchability). Hence the bias.

Illustrative simulations

The above example highlights the problem, but is not on its own evidence of bias due to heterogeneity. We can investigate this bias by simulation, using the library WiSP.

1,000 surveys were simulated for the homogeneous and heterogeneous catchability scenarios. Surveys in which $n_2 \geq n_1$ were excluded from the results presented, because they give infinite abundance estimates. (If $n_2 \geq n_1$ the estimator of Equation (5.3) is negative, but clearly $n_2 \geq n_1$ implies N is large; in the library WiSP, $\hat{N}$ is constrained to be positive and $n_2 \geq n_1$ results in $\hat{N} \to \infty$.) This is an undesirable property of the removal method estimator.

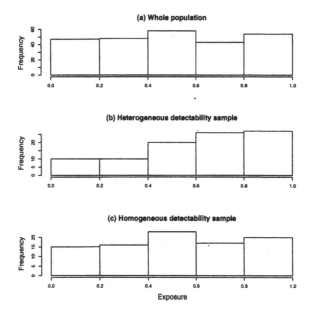

Figure 5.8. Distribution of animal exposure in (a) the whole population, (b) detected animals when detection probability depends on exposure, and (c) detected animals when detection probability does not depend on exposure.

With the heterogeneous catchability simulations, 2% of the estimators had $n_2 \geq n_1$; with homogeneous catchability, the figure was 16%. Excluding these cases, the mean MLE from the simulations is about 34% below the true N when there was heterogeneity, and 2% above when there was none. When there was heterogeneity, 90% of the $\hat{N}$s were below the true population size; without heterogeneity, the figure was 60% (or around 50% if the infinite abundance estimates are included). If we repeat the simulations with a less heterogeneous population, the negative bias decreases. The simulated distributions of the estimators are shown in Figure 5.9.

If heterogeneous catchability exists and is ignored, estimates of N are negatively biased, and this bias can be severe if there is substantial heterogeneity. Seber (1982) summarizes some analytic results on the size of bias due to changes in catchability. The R library WiSP can be used to investigate bias further, by simulation.

Simulations can also be used to investigate other properties of the estimator. For example, Figure 5.9 shows that the distribution of $\hat{N}$ is not close to normal, so one can expect CIs based on a normal approximation to perform poorly.

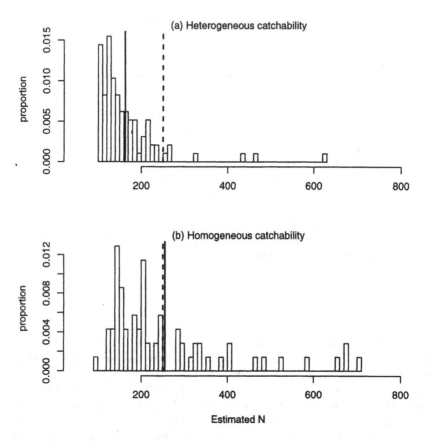

Figure 5.9. Simulated distribution of the two-sample removal method MLE for a simulated population when (a) catchability is heterogeneous and (b) catchability is homogeneous (see text for details). The solid vertical line is at the simulated $E[\hat{N}]$; the dotted vertical line is at the true N.

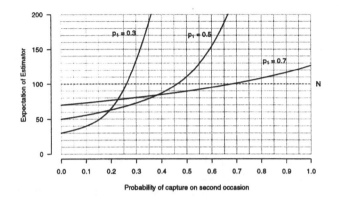

Figure 5.10. Expected value of $\hat{N}$ from the two-sample simple removal method in which the detection probability on the second occasion (p_2) differs from that on the first occasion (p_1) and inadmissable estimates are discarded, when $N = 100$ and $p = 0.3, 0.5, 0.7$.

Model mis-specification

The bias problem is due to model mis-specification: a model with constant p is inappropriate if p really depends on an animal-level variable x. Similar model mis-specification arises if p changes between surveys due to survey-level variables.

The following example illustrates the direction and severity of the bias that can result in a two-sample survey if the assumption that $p_1 = p_2$ is violated. Suppose that the detection probability on occasion 2 is p_2, which might differ from p_1, and that we estimate N using Equation (5.3), which is based on the assumption that p_1 and p_2 are equal. As already noted, this estimator is inadmissible whenever $n_2 \geq n_1$. In the computations that follow, we deal with this complication by discarding surveys in which such observations occur. That is, we compute $E[\hat{N}]$ given that $n_2 < n_1$.

Figure 5.10 shows plots of $E[\hat{N}]$ for $N = 100$, for $p_1 = 0.3, 0.5, 0.7$, and for a range of values of p_2. The plots illustrate that $E[\hat{N}]$ is biased even when $p_1 = p_2$, that is even when the assumption is met. The bias is negligible for $p = 0.7$, but it increases when p is decreased to 0.5 and then 0.3. Note also that if $p_2 < p_1$, we tend to underestimate N, and if $p_2 > p_1$, we tend to overestimate N. The bias resulting from violations of the assumption $p_1 = p_2$ becomes increasingly severe as p_1 decreases. Note how steeply $E[\hat{N}]$ increases with increasing p_2 for the case $p_1 = 0.3$, especially for $p_2 > p_1$. Finally, the figures illustrate that the bias increases more rapidly for $p_2 > p_1$ than it does for $p_2 < p_1$.

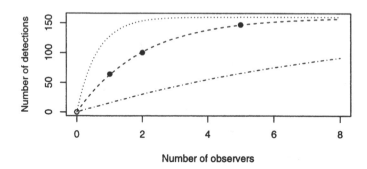

Figure 5.11. The number of animals detected by three observer teams, against size of the team (circles). The fitted detection function is shown (closest to the points), together with two other examples of the shapes of detection functions with the same form but different parameter values.

In the next two sections, we deal with models that allow p to depend on a survey-level variable. Models that allow p to depend on animal-level variables are substantially more complicated because we don't know the animal-level variables for undetected animals. In Chapter 11 we discuss fully model-based and partially design-based methods of dealing with this difficulty.

5.3   Catch-effort

If more effort is put into catching animals on some occasions, the assumption of constant p may be unreasonable. We would expect to catch a higher proportion of the population, on average, when we use more effort. The catch-effort method for closed populations is a simple extension of the removal method to reflect this expectation. Detection probability in each survey is assumed to depend on a measure of effort used in the survey.

We can illustrate the method with our example data by treating observer 1 as the first survey occasion (with one person-unit of effort), observers 2 and 3 combined as the second survey occasion (with two person-units of effort), and observers 4 to 8 as a third occasion (with five person-units of effort). We use l_s to indicate the person-units of effort used on occasion s, so that $l_1 = 1$, $l_2 = 2$ and $l_3 = 5$.

We expect two observers to be better at catching animals than one observer, and five to be better still. It is clear that this is the case here, as can be seen from Figure 5.11, which shows the number of animals detected by each of the three teams plotted against the size of the team (l_s) – if no animals were removed by any team.

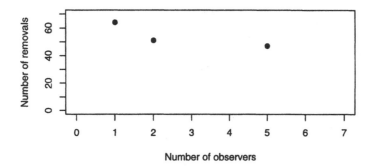

Figure 5.12. The number of animals detected by three observer teams, against size of the team, when teams remove all animals they detect.

The assumption of equal detection probability that is made for the simple removal method is clearly violated. But how do we use our knowledge of the effort employed on each occasion? To do this we need to model detection probability as a function of effort, i.e. we need to specify how it is affected by effort. We don't know what this relationship really is, so we use a functional form that can take on a wide range of sensible shapes, and let the data tell us which one of these shapes fits best. We'd want this function to be one which allows detection probability to go to zero as effort goes to zero (i.e. to go through the empty circle in Figure 5.11), to become one (certain detection) with enough observers, and to pass close to all the points in Figure 5.11.

The functional form $p(l_s) = 1 - \exp\{-\theta l_s\}$, where l_s is the number of observers on occasion s, is a suitable candidate. Figure 5.11 shows the sorts of shape the function takes, together with the function fitted to the example data. There are many other reasonable forms. Because we don't know exactly how effort affects detection probability, this function has unknown parameters, $\underline{\theta}$, which are estimated by fitting the function to the data (by maximizing the appropriate likelihood, for example).

The data and method we have described so far correspond to a "detection-effort" method, because no catches (removals) occurred. In practice this method is very rare; the common method involves removing all detected animals (by fishing for consumption, hunting, etc.). If we make the three observer teams in our example remove the animals they detect, the number of detections of the remaining animals looks very different from Figure 5.11 when plotted against team size. It is shown in Figure 5.12; despite at least doubling effort on each occasion, the number of detections falls from one occasion to the next. The "catch-per-unit-effort" (CPUE for short) drops dramatically between occasions, as shown in Figure 5.13. The obvious interpretation is that the population size has dropped substantially between

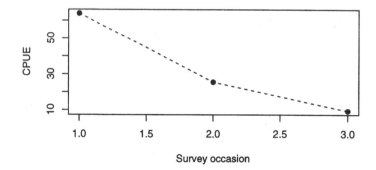

Figure 5.13. The catch-per-unit-effort, or "CPUE" (n_s/l_s) of the example population for each occasion.

occasions – that a substantial proportion of the remaining population has been removed on each occasion.

The relationship between detection probability and effort is no longer clear visually because of the removals. On the other hand, known removals provide additional information on abundance.

In the section below, we derive a likelihood for this scenario. By maximizing it, we obtain estimates of the parameters $\underline{\theta}$ (and hence estimates of detection probabilities) and N.

5.3.1 Likelihood

Technical section ↓

The method is based on the same assumptions as are used for the removal method, but allowing detection probability on each survey to depend on a measure of the effort used on that occasion. This might be the number of traps used, the time spent searching, or some other appropriate measure. The likelihood function is a simple extension of that for the removal method, with constant p replaced by $p(l_s)$ – a function of the effort, l_s.

$$L(N, \underline{\theta} \,|\, \{n, R\}) \;=\; \prod_{s=1}^{S} \binom{N_s}{n_s} p(l_s)^{n_s} \,(1 - p(l_s))^{N_s - n_s}$$

$$(5.10)$$

where $\{n, R\}$ and N_s are as for the removal method, and $\underline{\theta}$ is the unknown parameter vector of $p(l_s)$ (written θ when it is a scalar).

Suitable functional forms for $p(l_s)$ depend on the application and the assumptions made about the detection process. For example,

if l_s is the number of traps on survey s and each animal is assumed to be caught in any trap with probability θ, independently
of detections of other animals, then $p(l_s) = 1 - (1-\theta)^{l_s}$ is appropriate (see Exercise 5.5). In fisheries applications, where l_s might
be time spent fishing, $p(l_s) = 1 - \exp\{-\theta l_s\}$ has often been used,
as has the corresponding linear approximation $p(l_s) \approx \theta l_s$ (see Exercise 5.6), which is reasonable when θl_s (and hence p) is small.
Figure 5.11 shows three detection functions of this form. Another
general-purpose detection function form is the logistic; see Equation (B.7).

$\uparrow$
Technical
section

   With reasonable forms for $p(l_s)$, maximum likelihood estimates cannot
usually be obtained analytically. Nowadays this is not a real obstacle as
desktop computers have sufficient power to maximize the likelihoods in very
little time. (Nevertheless, unless there are many occasions and/or a large
proportion of the population is removed, maximizing the likelihood can be
difficult and maximization routines can fail – essentially because there are
insufficient data to fit the model reliably.) Before such computing power was
widely available, various computationally feasible regression methods were
developed as alternatives to maximum likelihood estimation. Most rely on
approximating the detection function by the linear function $p(l_s) \approx \theta l_s$.
This is reasonable provided $p(l_s)$ is small. We do not cover these methods
here; Seber (1982) covers a number of them.

5.3.2   Removal models and model selection

In going from the simple removal method to the catch-effort method, we
have accommodated changes in detection probability due to quantifiable
survey-level variables (the number of observers in our example). To do this,
we parametrized detection probability in terms of the survey-level variable,
$p(l_s; \underline{\theta})$, and we estimated $\underline{\theta}$ instead of estimating a constant p. One could
equally well use another sort of survey-level variable in this way, or use more
than one. In this case one needs to consider whether detection probability
really does depend on the variable(s) under consideration. This is a model
selection problem and it can be addressed using AIC. We illustrate with a
simple example.

   Using the R library WiSP, we generated a homogeneous population of $N = 1,000$ animals and conducted a catch-effort survey of it with five occasions.
True detection probability depended on effort as follows

$$p(l_s) = 1 - e^{-\theta \times l_s} \tag{5.11}$$

and effort increased as follows: $l_1 = 5$, $l_2 = 5$, $l_3 = 6$, $l_4 = 7$, $l_5 = 7.5$.
The survey data are summarized in Table 5.2. We estimated abundance

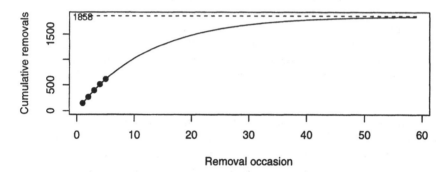

Figure 5.14. The expected and observed cumulative removals from the removal method survey of a homogeneous population of 1,000 animals. Estimated abundance is indicated by the dashed line.

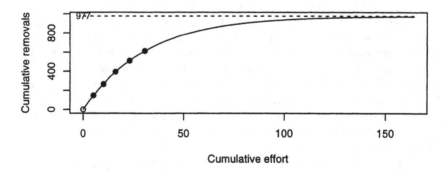

Figure 5.15. The expected and observed cumulative removals from the catch-effort method survey of a homogeneous population of 1,000 animals. Estimated abundance is indicated by the dashed line.

Table 5.2. Simulated survey of homogeneous population with $N = 1,000$.

	Survey occasion				
	1	2	3	4	5
Number of detections	148	118	128	117	101
Cumulative removals	0	148	266	394	511

Table 5.3. Simulated survey of homogeneous population with $N = 1,000$: log–likelihood, AIC, and estimated abundance for removal method and catch-effort method.

	Model	
Statistic	Removal	Catch-effort
log-likelihood	-17.5	-16.3
AIC	37.1	34.7
$\hat{N}$	1,858	977

using both the simple removal method and a catch-effort model with detection probability parametrized as above. Table 5.3 shows the maximum log-likelihood values, AIC, and estimated abundance $\hat{N}$ from both models

Reassuringly, the catch-effort model fits the data better (it has higher log-likelihood) and is the chosen model on the basis of its lower AIC. The removal method estimate is nearly twice the true abundance while the catch-effort estimate is extremely close to the true abundance. The removal method estimate is biased because it ignores the increasing trend in effort; it interprets the steadily increasing cumulative removals to mean that removals have reduced the population by much less than they have in reality (compare Figures 5.14 and 5.15). A related problem has been the downfall of many an exploited animal population. We discuss this briefly in the next subsection.

5.3.3 A word on CPUE as an index of abundance

Although we do not cover open populations in this book, a brief comment on the interpretation of CPUE plots and unmodelled survey-level variables for open populations seems in order because of its sometimes disastrous effect on the populations. There are many examples in the history of managing exploited populations in which a population collapse has been masked by technological improvements that increase the efficiency of catching. If catchers become more efficient (by technological innovation or otherwise), the CPUE can remain unchanged when the population is declining, because the drop in the number of animals available to be caught is compensated for by an increase in the probability of catching animals. The efficiency of catch technology is a relevant survey-level variable, but it is notoriously difficult to quantify. Unmodelled catchability due to changes in survey-level variables can result in bias that renders the catch-effort method virtually useless.

5.4   Change-in-ratio

The catch-effort method is designed to cope with change in detection probability due to survey-level variables (effort in particular), but it takes no account of differences in animal-level variables (size, sex, etc.). By contrast, the change-in-ratio (CIR) method depends on the presence of observable differences in animal-level variables. In particular, it relies on

(1) the relevant animal-level variables taking on one of a finite number of known values (most often just one of two values), and

(2) detection probability not depending on animal-level variables.

Although it does involve animal-level variables, the simple change-in-ratio method is not a method that accommodates heterogeneity in detection probabilities due to animal-level variables.

Suppose that we survey a population, and detect 16 females and 48 males. Assuming that males and females are equally detectable, we estimate that about $48/(48 + 16) = 3/4$ of the population are male. Suppose that we remove the 48 detected males from the population, and conduct a second survey. If we now detect say 20 females and 20 males, we estimate that the removal of 48 males was sufficient to reduce the proportion of males from 0.75 to 0.5. Hence if N is the initial population size, comprising N_f females and N_m males, we estimate that $N_m = 3N_f$ and $N_m - 48 = N_f$. Solving these equations yields $N_f = 24$, $N_m = 72$ and $N = 96$.

Note two things about the rationale for the method:

(1) We did not make any assumptions about the relationship between the detection probabilities on the two occasions. All we assumed was that the two types of animal were equally catchable. The rationale applies irrespective of whether detection probability on occasion 1 is equal to, less than, or greater than that on occasion 2.

(2) The rationale did not require us to measure or know the survey effort applied on each occasion.

One might expect that if we did know something about the survey effort and detection probabilities applied on each occasion (as with the catch-effort method), this might improve estimation in some way. We return to this point below.

We can generalize our estimator of abundance for the two-sample change-in-ratio method as follows. Let $N_s(x)$ be the number of animals of sex x in the population at the start of survey occasion s, and N_s be the population size at the start of occasion s. Similarly, let $R_2(x)$ be the number of animals of sex x that are removed by the start of survey occasion 2, and R_2 be the total removed by the start of occasion 2. Then the proportion of males in the population on the second occasion $(\pi_2(1) = N_2(1)/N_2)$ can be written as

$$\pi_2(1) \quad = \quad \frac{\pi_1(1)N_1 - R_2(1)}{N_1 - R_2} \tag{5.12}$$

and solving for N_1 (which is also N, the initial population size), we get

$$N \quad = \quad \frac{R_2(1) - R_2\pi_2(1)}{\pi_1(1) - \pi_2(1)} \tag{5.13}$$

where $\pi_1(1) = N_1(1)/N_1$ is the initial proportion of males in the population. We don't know $\pi_1(1)$ or $\pi_2(1)$. But if males and females are equally catchable, we can reasonably estimate them using the proportion of males in the sample on each occasion:

$$\hat{\pi}_1(1) \quad = \quad \frac{n_1(1)}{n_1} \qquad\qquad \hat{\pi}_2(1) \quad = \quad \frac{n_2(1)}{n_2} \tag{5.14}$$

where n_1 and n_2 are the sample sizes on occasions 1 and 2. Using these in Equation (5.13), we get our estimator:

$$\hat{N} \quad = \quad \frac{R_2(1) - R_2\hat{\pi}_2(1)}{\hat{\pi}_1(1) - \hat{\pi}_2(1)} \tag{5.15}$$

We use our example data for illustration, considering observer 1 to be the first survey occasion and observers 2 and 3 to be the second occasion – as we did for the catch-effort method above. On the first survey of our example dataset, $n_1(0) = 16$ females and $n_1(1) = 48$ males were detected. Suppose the 48 males were then removed ($R_2(0) = 0$; $R_2(1) = 48$). On the second survey, $n_2(0) = 26$ females and $n_2(1) = 38$ previously undetected males were detected.

With these data, we estimate the initial proportion of males to be $\hat{\pi}_1(1) = 48/64$, and the proportion on occasion 2 to be $\hat{\pi}_2(1) = 38/64$. Removing 48 males appears to have reduced the proportion of males quite substantially (from 0.75 to about 0.6), which suggests that the initial number of males in the population was not vastly more than the number removed. Using Equation (5.15), we estimate abundance to be

$$\hat{N} = \frac{48 - 48 \times \frac{38}{64}}{\frac{48}{64} - \frac{38}{64}} \approx 125 \qquad (5.16)$$

($\hat{N}$ is half the true population size!). We estimate the number of males initially in the population to be $\hat{\pi}_1(1) \times 125 \approx 94$.

Variance can be estimated by

$$\widehat{Var}[\hat{N}] = \frac{\sum_{s=1}^{2} \hat{N}_s^2 \widehat{Var}[\hat{\pi}_s(1)]}{(\hat{\pi}_1(1) - \hat{\pi}_2(1))^2}$$

$$\approx 1,087 \qquad (5.17)$$

(an estimator from Seber, 1982, pp 355–356), where

$$\widehat{Var}[\hat{\pi}_s(1)] = \frac{\hat{\pi}_s(1)(1 - \hat{\pi}_s(1))}{(n_s - 1)} \left[1 - \frac{n_s}{\hat{N}_s}\right] \qquad (5.18)$$

for $s = 1, 2$. (This variance estimator is based on the product hypergeometric conditional likelihood, given n_1 and n_2 – covered in Section 5.4.2 below. An alternative estimator, based on slightly different assumptions, is $\widehat{Var}[\hat{\pi}_s(1)] = \hat{\pi}_s(1)(1 - \hat{\pi}_s(1))/n_s$; see Exercise 5.7.)

Assuming that $\hat{N}$ is normally distributed, we estimate the 95% confidence interval for N to be $(60; 189)$. Again we have a confidence interval that does not include the true $N = 250$.

Although we expect profile likelihood or nonparametric bootstrap confidence intervals to be better than the above interval based on asymptotic normality, we know this population is heterogeneous, and for the same reasons that this results in negative bias in $\hat{N}$ in the case of the removal method, it will generate negative bias in $\hat{N}$ from the change-in-ratio method.

5.4.1 Full likelihood

Technical
section
↓

The number of animals of each of the K types removed from the population by occasion s is given by the vector $\underline{R}_s$, while the number of animals of each of the K types remaining in the population on occasion s is given by the vector $\underline{N}_s$:

$$\underline{N}_s \quad = \quad \underline{N}_1 \quad - \quad \underline{R}_s$$

or

$$\begin{pmatrix} N_s(\underline{x}_1) \\ \vdots \\ N_s(\underline{x}_K) \end{pmatrix} = \begin{pmatrix} N_1(\underline{x}_1) \\ \vdots \\ N_1(\underline{x}_K) \end{pmatrix} - \begin{pmatrix} R_s(\underline{x}_1) \\ \vdots \\ R_s(\underline{x}_K) \end{pmatrix}$$

Similarly, $\underline{n}_s$ is the "observation vector" giving the number of animals of each of the K possible types in the population which are observed on survey occasion s:

$$\underline{n}_s \quad = \quad \begin{pmatrix} n_s(\underline{x}_1) \\ n_s(\underline{x}_2) \\ \vdots \\ n_s(\underline{x}_K) \end{pmatrix} \tag{5.19}$$

With this notation, the full likelihood for the change-in-ratio method can be written as

$$L(N, \underline{\theta} \,|\, \{\underline{n}, \underline{R}\}) \quad = \quad \prod_{s=1}^{S} L(N, p \,|\, \underline{n}_s, \underline{R}_s) \tag{5.20}$$

where

$$L(N, p \,|\, \underline{n}_s, \underline{R}_s) \quad = \quad \prod_{\underline{x}} \binom{N_s(\underline{x})}{n_s(\underline{x})} p(l_s)^{n_s(\underline{x})} (1 - p(l_s))^{N_s(\underline{x}) - n_s(\underline{x})}$$

$$\tag{5.21}$$

↑
Technical
section

5.4.2 Conditional likelihood

The likelihood Equation (5.20) is appropriate if animals are detected on each occasion (n_s, for $s = 1, \ldots, S$) with probability $p(l_s)$. In many applications of the CIR method, animals are either detected by some unknown means, or the numbers detected on each occasion are specified in advance. In both these cases, the n_ss contain no useful information about N and it is appropriate to treat the n_ss as given (not as random variables), and to work with the conditional likelihood given n_s, for $s = 1, \ldots, S$.

Technical
section
↓

Equation (5.20) can be factorized as follows:

$$
\begin{aligned}
L &= \left[\prod_{s=1}^{S} \binom{N_s}{n_s} p(l_s)^{n_s} \left[1 - p(l_s) \right]^{N_s - n_s} \right] \times \left[\prod_{s=1}^{S} \frac{\prod_{\underline{x}} \binom{N_s(\underline{x})}{n(\underline{x})}}{\binom{N_s}{n_s}} \right] \\
&= \left[\prod_{s=1}^{S} P(n_s | N_s) \right] \times \left[\prod_{s=1}^{S} P(\underline{n}_s | n_s, \underline{N}_s) \right]
\end{aligned}
\qquad (5.22)
$$

where $n_s = \sum_{\underline{x}} n_s(\underline{x})$, $N_s = \sum_{\underline{x}} N_s(\underline{x})$,

The second term on the right, $\prod_{s=1}^{S} P(\underline{n}_s | n_s, \underline{N}_s)$ is the conditional probability of observing $\underline{n}_s$, given n_s and $\underline{N}_s$. When $\underline{x}$ is a scalar, that can take on only two values (e.g. $x = 0$ for females, $x = 1$ for males), this is the product of hypergeometric densities which is usually associated with the CIR method (e.g. Seber, 1982; page 356). When $\underline{x}$ can take on more than two values, the appropriate multivariate hypergeometric replaces the hypergeometric distribution in the second term. Considered as a function of N, it is the conditional likelihood of N, given n_s $(s = 1, \ldots, S)$.

With only two survey occasions $(S = 2)$, Equation (5.15) is the conditional MLE.

The additional leading term, $\prod_{s=1}^{S} P(n_s | N_s)$, generalizes Seber's likelihood somewhat. Because the second component of the full likelihood conditions on the number of animals observed $(n_s; s = 1, \ldots, S)$, it contains no data on the detectability of animals $(p(l_s)$ above). This is appropriate if the n_ss are not random variables, but when they are, the leading term above allows these data to be incorporated in estimation. Chapman and Murphy (1965) dealt slightly differently with a two-sample CIR method in which n_1 and n_2 are random variables — by assuming that $n_s(x)$ $(s = 1, 2;$ $x = 0, 1)$ are independent Poisson random variables.

↑
Technica
section

As with the simple removal method and the catch-effort method, neglecting heterogeneity in the case of the change-in-ratio method results in biased estimation, and this bias can be large. Moreover, unless the ratio of animal types is changed substantially by removal, the method will frequently give very unreliable or simply inadmissible estimates of abundance. This should be clear from Equation (5.15) – if the true proportion of males in the population has changed little, the probability of the observed proportion being equal to or larger on the second occasion than on the first will be high; hence the probability that the estimate of N is inadmissible will be high.

Using the full likelihood instead of the conditional likelihood should help as it includes catch-effort information as well as information about the change in ratio. Even in this case, however, we have found that when the number of occasions is small and/or a small fraction of the population is removed, the frequency with which numerical routines fail to maximize the likelihood can be large.

5.5   Effect of violating main assumptions

Assumption 1a (simple removal method). All animals are detected with equal probability, p.

Effect of violation: Estimates of abundance are negatively biased by violation of this assumption. The more p varies in the population (the more heterogeneity), the larger the bias. It can be large. The simple removal method is not robust to "pooling" animals with different detection probabilities.

Changes in p between surveys due to survey-level variability can also cause large bias (witness the example with increasing effort, above). The catch-effort method tries to deal with this problem.

Assumption 1b (catch-effort method). Detection probability depends only on survey-level variables, like measured effort.

Effect of violation: Estimates of abundance are negatively biased if detection probability depends on animal-level variables, as it does in many cases.

Assumption 1c (change-in-ratio method). Detection probability may vary between occasions, but is constant for all animals within occasions.

Effect of violation: Estimates of abundance are negatively biased if detection probability depends on animal-level variables – as it does in many cases.

Assumption 2 Removals are known.

Effect of violation: Bias and/or variance inflation. There are methods that try to deal with uncertainty in removals; we do not deal with them in this book.

5.6   Summary

None of the methods of this chapter involve a state model and all uncertainty arises from the observation model – the fact that not all animals present on each occasion are observed. The covered region is assumed to be the survey region and all animals in the population are assumed to be subject to the same effort on any given occasion. While the change-in-ratio method involves x, it involves only the sort of xs that can take on a few discrete values and the numbers of animals with each x-value (number of males and number of females, for example) are treated as fixed parameters, not realizations of an underlying state model for x. In Part III, we reconsider removal methods with heterogeneity and state models.

The three sorts of removal method (simple removal, catch-effort, and change-in-ratio) are really special cases of the same general method (corresponding to the likelihood function in the box at the start of the chapter). This fact was recognized by Udevitz and Pollock (1995) and by Chen et al. (1998) who developed likelihood functions and estimators that combine two or more of the three sorts of removal method. They used a slightly different likelihood formulation than is presented here; the main difference being that they assume Poisson counts rather than binomial counts, and they use different parametrizations of the capture functions.

The simple removal method recognizes neither different kinds of animal nor different levels of effort (it involves neither x nor l_s). The catch-effort method accommodates different levels of effort on different occasions (detection probability is a function of effort, l_s), but treats all animals as equally catchable on any given occasion. The change-in-ratio method is the most general of the three in that it involves different kinds of animal (different x) and accommodates different levels of effort on different occasions. None of the methods accommodate heterogeneity (detection probability depending on animal-level variables, x) and all are subject to potentially large bias if heterogeneity exists. Dealing with heterogeneity appropriately is clearly important for unbiased estimation. We discuss methods that do this in Chapter 11.

To recap: with plot survey methods, detection probability is known; with removal methods it is not. The change in detection rate after the removal of some animals provides the crucial data that enables these methods to estimate abundance without knowing detection probability. Aside from their physical cost to the population, the cost of removals to the estimation process is that data from the removed animals contribute nothing once the animals are removed. In the following chapter, we consider methods in which animals are marked rather than removed. Marking provides the same sort of data as removal (unmarked animals correspond to unremoved animals) without losing animals from the population. Recaptures of marked animals contribute a new kind of data that supplements the "removal" data.

5.7   Exercises

Exercise 5.1   Derive the maximum likelihood estimators of N and p given in Equation (5.5), from the likelihood Equation (5.4).

Exercise 5.2   Hence show that the MLEs for the two-sample simple removal method are given by Equations (5.3) and (5.6).

Exercise 5.3   Consider a two-sample simple removal method in which detection probability is assumed to be constant at p on both occasions, but in fact changes between occasions. Show that the most extreme bias that can result if $p_2 < p$ is $E[\hat{N}] = Np$.

Exercise 5.4   Show that given N, the two-sample simple removal method likelihood is maximized at p equal to

$$\hat{p}(N) = \frac{n_1 + n_2}{2N - n_1} \tag{5.23}$$

Exercise 5.5   The catch-effort method is used to estimate the abundance of a closed population by setting l_s baited traps on occasion s ($s = 1, \ldots, S$). Assume that every animal has the same probability θ of being captured in each trap on each occasion, that animals are captured independently, and that capture probability is not changed if a trap already has one or more occupants. By considering the probability that on any one occasion, an animal is not captured in any of the l_s traps, show that the probability of catching an animal when l_s traps are used is $1 - (1 - \theta)^{l_s}$.

Exercise 5.6   Suppose that the probability of catching a fish with l_s units of fishing effort is $p(l_s) = 1 - \exp(-\theta l_s)$.

 (a) By considering a Taylor series expansion of this function, show that when $p(l_s)$ is small, $p(l_s) \approx \theta l_s$.

 (b) If $p(l_s) = \theta l_s$, show that the expected CPUE on occasion s is $\theta \times (N - R_s)$.

 (c) Figure 5.16 shows the CPUE of a fish population in a lake, plotted against R_s. If $p(l_s) = \theta l_s$, estimate θ visually from the plot. Hence estimate N.

Exercise 5.7   Suppose you conduct a two-sample change-in-ratio experiment on a population of $N = N_1$ animals consisting of $N_1(0)$ females and

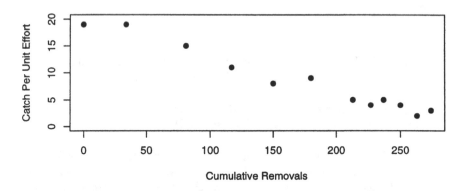

Figure 5.16. Plot of catch per unit effort (CPUE) against cumulative catch (R_s) for a population of fish in a lake.

$N_1(1)$ males at the start of the first survey occasion. $R_2(0)$ females and $R_2(1)$ males are removed before the second survey occasion.

(a) A total of n_1 animals are observed on the first survey occasion while n_2 are observed on the second. Show that the probability of observing $n_1(0)$ females on the first occasion and $n_2(0)$ females on the second, given n_1 and n_2, is the following product of hypergeometric probability mass functions:

$$P(\underline{n}(0)|\underline{n}) \quad = \quad \prod_{s=1}^{2} \frac{\binom{N_s(0)}{n_s(0)}\binom{N_s - N_s(0)}{n_s - n_s(0)}}{\binom{N_s}{n_s}} \qquad (5.24)$$

where $\underline{n}(0) = (n_1(0), n_2(0))'$, $\underline{n} = (n_1, n_2)'$, $N_s(0) = N_1 - R_s(0)$ and $N_s = N_1 - R_s = N_1 - [R_s(0) + R_s(1)]$.

(b) Treating Equation (5.24) as a likelihood for N and $N_1(0)$, show that Equation (5.15) is the MLE of N.

(c) Suppose that sampling was with replacement. If the probability of detecting any animal (male or female) is p, show the following:

  (i) The probability mass function of $n_s(0)$, the number of females detected on occasion s, is binomial, with parameters N and $p\pi_s(0)$, where $\pi_s(0)$ is the proportion of animals in the population on occasion s that are female.

  (ii) The conditional probability that an animal is female, given that it was detected, is $\pi_s(0)$.

(iii) Given that n_s animals were detected on occasion s, the probability mass function of $n_s(0)$, the number of females detected on occasion s, is binomial, with parameters n_s and $\pi_s(0)$.

(iv) Treating the probability mass function of (iii) above as a likelihood for $\pi_s(0)$, show that the MLE of $\phi_s(0)$ is $\hat{\pi}_s(0) = n_s(0)/n_s$ and that

$$Var[\hat{\pi}_s(0)] = \frac{n_s(0)n_s(1)}{n_s} \qquad (5.25)$$

Exercise 5.8   The abundance and detection probability estimated by the simple removal method from a two–sample removal method survey of the St Andrews example population are $\hat{N} = 132$ and $\hat{p} = 0.48$.

(a) What is the estimated probability that an animal survives these two removal occasions?

(b) What is the estimated probability that an animal survives the first two removal occasions and is caught on the third?

(c) What are the predicted removals on a third removal occasion?

(d) How many removal occasions are required until the expected number of removals on a single occasion falls below one?

6
Simple mark-recapture

Key idea: estimate p from proportion of marked animals recaptured.

Number of surveys:	At least two surveys required.
State model:	None. Assume complete coverage of the survey region.
Observation models:	Various. With simplest, capture probability is the same for all animals on all occasions. Other models allow it to depend on capture occasion and/or trap response. Independent captures.

6.1 Introduction

With data from one survey only, we can't estimate abundance without making strong assumptions about capture probability. In the case of plot surveys, we assume we know it, and with distance sampling methods (Chapter 7) we assume "capture" is certain on the line or point. Depending on the

---

Likelihood function and key notation:

$$L(\underline{N}_1, \underline{\theta}) = \prod_{s=1}^{S} \prod_{\underline{\omega}_s} \binom{N_s(\underline{x}_s)}{n_s(\underline{x}_s)} p(\underline{x}_s, s)^{n_s(\underline{x}_s)} (1 - p(\underline{x}_s, s))^{N_s(\underline{x}_s) - n_s(\underline{x}_s)}$$

$\underline{x}_s$:   animal "type", which is mark indicator or capture history or capture history and individual identity (see text).

$N_s(\underline{x}_s)$:   abundance of animals of type $\underline{x}_s$ on occasion s.

$n_s(\underline{x}_s)$:   number of animals of type $\underline{x}_s$ captured on occasion s.

$p(\underline{x}_s, s)$:   probability of catching an animal of type $\underline{x}_s$ on survey s.

$\underline{\theta}$:   parameters of the detection or capture function, $p(\underline{x}_s, s)$.

---

application, these assumptions may be quite reasonable; when they are not, removal or mark-recapture methods can be useful. With mark-recapture methods, animals continue to contribute information about capture probability after they have been initially captured; with removal methods, they do not. Recapture data allow us to estimate capture probability without the assumptions of plot or distance sampling methods and without removing animals from the population.

The basis of the methods is really quite simple. If the animals we capture on one occasion are typical of the animals in the population, they constitute a representative sub-population of known size. The proportion of this marked sub-population that we capture on a subsequent capture occasion is an obvious estimator of the probability of catching an animal, p. Applying this estimate to the whole population, we can estimate abundance, N. You can see that the methods rely heavily on the captured sub-population being representative of the whole population, and this can be a problem.

Animals can be captured and marked in a variety of ways. Capture might be physical capture or simply detection (by eye, satellite, radio, or other means). Marking might involve physically attaching a mark to the animal, recording a natural marking (tigers' stripes or humpback whales' flukes, for example), or notional marking of the animal by its location at a given time (as with the St Andrews example dataset).

Different kinds of marks contain different kinds of information:

(1) Some marks contain no information on previous capture history, but simply indicate that the animal has been caught at least once (this would be the case if marks were just a red spot painted on captured animals).

(2) Some marks tell you when the animal was captured (this would be the case if a different colour was used on each survey occasion).

(3) Some marks allow individuals to be recognized (this would be the case if a different colour was used on each survey occasion, together with a unique animal number for every captured animal).

Ideally all marks would be of type (3) because they contain more information useful for estimating abundance and allow a wider range of analyses to be conducted. Practical considerations often preclude the use of this sort of mark, however.

More sophisticated analyses are possible with type (2) than type (1), and with type (3) than type (2). In the interests of keeping things relatively simple, we focus on the analysis of marks of type (1) in this chapter. We mention the analysis of the other two types only briefly.

6.2 Single recapture and some notation

Although the likelihood function above is formulated for marks of type (1), (2) or (3), we simplify our notation for most of this chapter in line with our focus on the analysis of mark type (1) data. We drop the bracketed arguments for Ns and ns, and we use a simplified notation for ps. We continue to use uppercase letters to indicate numbers in the population and lowercase letters to indicate numbers in the samples. The notation is illustrated in Figure 6.1.

Abundances The number of animals in the population at the start of occasion s is N_s. Of these, $N_s(1)$ are marked (i.e. have $x_s = 1$). For brevity, we use M_s instead of $N_s(1)$ to indicate the number of marked animals in the population at the start of occasion s. Similarly, we abbreviate $N_s(0)$ to U_s, the number of unmarked animals in the population at the start of occasion s: $M_s + U_s = N_s$.

Numbers captured The number of marked animals (with $x_s = 1$) at the start of occasion s that are captured on occasion s, is $n_s(1)$. For brevity, we often use m_s instead of $n_s(1)$. Similarly, we abbreviate the number of unmarked animals that are captured on occasion s, $n_s(0)$, to u_s: $m_s + u_s = n_s$.

Capture probabilities Our general notation for the probability of capturing an animal of type $\underline{x}$ on occasion s is $p(\underline{x}, s)$. We use a shorter notation in this chapter. When capture probability depends only on occasion, and not on animal type, we write this as p_s for brevity. When it depends only on whether or not an animal has been marked,

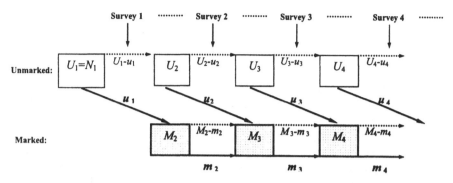

Figure 6.1. Schematic representation of a mark-recapture survey. M_s is the number of marked animals in the population at the start of survey occasion s; U_s is the number of unmarked animals in the population at the start of survey occasion s. m_s is the number of marked animals that are captured on survey occasion s; u_s is the number of unmarked animals that are captured on survey occasion s. Shaded boxes represent animals that have been captured at least once (so the numbers in the shaded boxes are known); unshaded boxes represent animals that have never been caught. Dark arrows represent the paths of animals that are captured in the survey shown; dotted arrows represent the paths of animals that are not captured.

we write it as p_u (for unmarked animals) and p_m (for marked animals). When it depends neither on animal type nor on occasion, we write it as p.

The simplest sort of mark-recapture experiment is one with a single marking occasion and a single recapture occasion in which all animals are equally likely to be captured. We use the St Andrews example dataset to illustrate it. In this case, detection of an animal constitutes a visual capture. Recaptures are animals detected by more than one observer.

6.2.1 St Andrews data example

At the start of the first survey there are no marked animals, so $M_1 = 0$. With our example data, $u_1 = 64$ animals were captured (visually) by Camilla, who did the first survey. By the time Ian did his survey (the second), 64 animals were already marked, so $M_2 = 64$. Of these, he recaptured $m_2 = 34$. Without any statistical development, we could reasonably estimate the probability that Ian captures an animal by the proportion of the marked animals that he captured: $\hat{p}_2 = m_2/M_2 = 34/64 = 0.53125$. (Note that we can also write this as m_2/u_1 or as m_2/n_1.) Since he caught a total of $n_2 = 67$ animals, we expect that 67 is about 53% of the population. Put another way: $\hat{N} \times 0.53125 = 67$. Rearranging this equation gives an estimate $\hat{N} = 126$. In general

$$\hat{N} \doteq \frac{n_2}{\hat{p}_2} = \frac{n_2}{\left(\frac{m_2}{n_1}\right)} = \frac{n_1 \times n_2}{m_2} \qquad (6.1)$$

(Recall that $n_1 = u_1 = M_2$ because the unmarked animals captured on occasion 1 are all marked.) This is the "Petersen mark-recapture estimator" or the "Lincoln index".

We could similarly estimate the probability that Camilla captures an animal as $\hat{p}_1 = m_2/n_2 = 34/67 \approx 0.507$. So with two surveys and data describing which animals were captured in which surveys, we are able to estimate capture probabilities (p_1 and p_2) and group abundance (N). Because marking on the first capture occasion creates a sub-population of $n_1 = u_1 = M_2$ marked animals, by the second capture occasion we are surveying a sub-population of known size (M_2) for which only one parameter is unknown (p_2). We get one relevant bit of data relating to this sub-population (namely m_2), and this is sufficient to estimate one parameter: $\hat{p}_2 = m_2/M_2$. Now if the marked and unmarked animals are equally detectable, then this estimate of p_2 applies to the whole population, not just the marked animals, and using it with all the data from capture occasion 2 (n_2), we are able to estimate N. (The "if" is sometimes a big "if".)

It turns out that the Petersen estimator is positively biased if the assumptions of the method hold, and this bias can be large for small samples. Chapman (1951) suggested a modified estimator

$$\hat{N} = \frac{(n_1 + 1)(n_2 + 1)}{(m_2 + 1)} - 1 \qquad (6.2)$$

which is approximately unbiased for $n_1 + n_2 < N$ and exactly unbiased if $n_1 + n_2 \geq N$. For this reason it is often used in preference to the Petersen estimator. In this particular example, the Petersen and Chapman estimators are virtually identical (126 and 125 respectively).

The estimate $\hat{N}$ is much less than the true N. There is a reason that this will on average be the case (for both Petersen's and Chapman's estimators) with these data (i.e. that $\hat{N}$ is negatively biased); the reason centres on the "if" above and has to do with how representative the captured animals are of the whole population. We discuss this further in Section 6.3.5.

6.3   A two-sample mark-recapture likelihood

Technical section
↓

As always, we need to specify our probability model precisely in order to get a likelihood function. In this chapter, we assume that all animals in the population are at risk of being caught, and capture probability does not depend on individual animal characteristics (we deal with more complicated models in Chapters 11 and 12). As a result we don't need a state model to draw inferences outside of our covered region; the covered region is the survey region and that is all we are interested in. The likelihood therefore consists only of an observation model (where "observation"=capture). In order to derive it, we start by assuming the following.

(1) **All animals are equally catchable on any one survey occasion**; we assume that capture probability can change between surveys: it is p_1 on survey 1 and p_2 on survey 2.

(2) **Detections of animals are independent events**, both within a survey and between surveys. That is, the fact that an animal is captured on a survey occasion does not affect the probability that any other animal is captured (within-survey independence), and the fact that an animal was captured on one occasion does not affect the probability that it is captured on another occasion (between–survey independence). Within-survey independence might not hold if one animal drew other animals to a trap, for example. Between-survey independence would not hold if an animal learned to avoid traps after being caught once, for example.

We now derive a likelihood for N, p_1 and p_2, given the number of marked animals in the population, by building on the results of Chapter 2.

6.3.1   First capture occasion

Because animals are assumed to be captured independently with equal probability p_1 on survey 1, the probability of capturing n_1 animals has the same form as Equation (2.3). That is, the probability of catching $n_1 = u_1$ unmarked animals on the first occasion, given that there are $U_1 = N$ animals in the population, is

$$P_1 \;=\; \binom{U_1}{u_1} p_1^{u_1} [1 - p_1]^{U_1 - u_1} \qquad (6.3)$$

Considered as a function of $N = U_1$ and p_1, this is the likelihood function for the first capture occasion.

6.3.2   Second capture occasion

By marking animals caught on the first survey, the population is split into two types by the time of the second survey: marked and unmarked. When all captured animals are marked on the first occasion, the number of marked animals by the start of the second occasion is the number of animals captured on the first occasion, i.e. $M_2 = n_1 = u_1$. Because we know how many marked animals there are by the start of the second survey, the outcome of this survey is not just the number captured on the survey (n_2), it is how many of the known M_2 marked animals were captured (namely, m_2) and how many of the unknown $U_2 = (N - M_2)$ unmarked animals were captured (namely, u_2). The probability of observing $\underline{n}_2 = (u_2, m_2)^T$ is therefore a product of two binomial likelihoods. The first is the probability of catching m_2 marked animals:

$$P_{2m} = \binom{M_2}{m_2} p_2^{m_2} [1 - p_2]^{M_2 - m_2} \qquad (6.4)$$

Given m_2, this is a likelihood function with only one unknown parameter, p_2 (remember that M_2 is known to be the number marked on occasion 1). With one bit of data on marked animals (m_2), we can estimate p_2: the MLE is $\hat{p}_2 = m_2/M_2$. In the case of our example data, this is 34/64.

There is a similar probability for unmarked animals, which is as follows:

$$P_{2u} = \binom{U_2}{u_2} p_2^{u_2} [1 - p_2]^{U_2 - u_2} \qquad (6.5)$$

Given u_2, this is a likelihood function with two unknown parameters, p_2 and U_2. With only one bit of data on unmarked animals (m_2), we can't estimate p_2 or U_2 from this likelihood alone. What we could do is to "plug in" the estimate of p_2 from the likelihood for marked animals, Equation (6.4), leaving only M_2 to be estimated from the likelihood for unmarked animals. This way we can estimate both p_2 and N (which is equal to $M_2 + U_2$).

Alternatively, we could combine Equations (6.4) and (6.5), and estimate p_2 and N in one fell swoop from the likelihood

$$L_{2 \text{ only}} = P_{2u} \times P_{2m} \qquad (6.6)$$

$$= \binom{U_2}{u_2} p_2^{u_2} [1 - p_2]^{U_2 - u_2} \binom{M_2}{m_2} p_2^{m_2} [1 - p_2]^{M_2 - m_2}$$

We can also write this in our general notation, using $\underline{x}_s$ to indicate animal "type" on occasion s (with $\underline{x}_2 = 1$ for marked animals and $\underline{x}_2 = 0$ for unmarked animals), as

$$L_{2 \text{ only}} = P_{2u} \times P_{2m} \tag{6.7}$$

$$= \prod_{x_2} \binom{N_2(x_2)}{n_2(x_2)} p(s=2)^{n_2(x_2)} [1 - p(s=2)]^{N_2(x_2) - n_2(x_2)}$$

where $\prod_{x_2}$ is the product over the two possible capture types (captured or uncaptured) by the start of occasion 2.

6.3.3 Putting the two capture occasions together

If we wanted to, we could plug the estimate of N from the second capture occasion into Equation (6.3) and estimate p_1 from this equation. Alternatively, we could estimate N, p_1 and p_2 simultaneously from the full likelihood for both capture occasions, given the observed data u_1, m_2 and u_2. The full likelihood can be written as

$$L_2 = \prod_{s=1}^{2} P_{su} \times P_{sm} \tag{6.8}$$

$$= \prod_{s=1}^{2} \binom{U_s}{u_s} p_s^{u_s} [1 - p_s]^{U_s - u_s} \binom{M_s}{m_s} p_s^{u_s} [1 - p_s]^{M_s - m_s}$$

(We define $P_{1m} \equiv 1$; this is a convenient device to allow us to use the product notation above for occasion $s = 1$.)

One can show that the likelihood Equation (6.8) is multinomial, with 4 cells, and that the Petersen estimator of N is the MLE (see Exercise 6.3).

↑
Technical
section

6.3.4 Interval estimation

Profile likelihood CI

As with the simple removal methods of the previous chapter, all uncertainty in simple mark-recapture estimates arises from the observation process; variance and confidence interval estimates are based on the observation model alone. As in the previous chapter, the observation model is the product of binomial pdfs (but not the same ones as in the previous chapter). Profile likelihood confidence intervals can be constructed using this pdf. For the case in which capture probability is the same on both capture

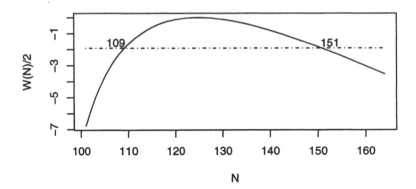

Figure 6.2. $W(N)/2 = l(\hat{N}, \hat{p}) - l(N, \hat{p}(N))$: the difference between the profile log-likelihood function and its maximum, for $100 \le N \le 165$, when $n_1 = 64$ captures on occasion 1, $n_2 = 67$ captures on occasion 2, and $m_2 = 34$ recaptures.

occasions, the profile likelihood is obtained by evaluating Equation (6.8) with $p_1 = p_2 = p$ at

$$p_1(N) \equiv p_2(N) \equiv p(N) \quad = \quad \frac{n_1 + n_2}{2N} \qquad (6.9)$$

(see Exercise 6.2). Figure 6.2 shows $W(N)/2 = l(\hat{N}, \hat{p}) - l(N, \hat{p}(N))$, the difference between the profile log-likelihood function and its maximum. For our example data, it gives a CI of (109, 151). This is worrying because we know in this case that $N = 250$.

Normal-based CI

An estimator of the asymptotic variance of the MLE for N can be obtained using the information matrix as outlined in Appendix C. Alternatively, we can use the following approximately unbiased estimator of the variance of Chapman's modified estimator, due to Seber (1970) and Wittes (1972).

$$Var[\hat{N}] \quad = \quad \frac{(n_1 + 1)(n_2 + 1)(n_1 - m_2)(n_2 - m_2)}{(m_2 + 1)^2(m_2 + 2)} \qquad (6.10)$$

Using this estimator, the normal-based CI is (105, 145), while the log-normal based CI is (107, 146).

Percentile method confidence interval

As with the removal methods considered in the previous chapter, the survey design is "cover the whole survey region" so there is only one designed sampling unit (the whole region) and a simple nonparametric bootstrap using this as the resampling unit is no use. The only source of variance is due to the randomness in the observation process. One way to capture this in a bootstrap procedure is to resample parametrically using our fitted model of the observation process, in much the same way as for the simple removal method, but sampling from the appropriate multinomial distribution instead of from a binomial. The method works by sampling from the appropriate multinomial pdf. With two capture occasions and p_1 not necessarily equal to p_2, this is a multinomial with parameters $\hat{N}$, $\hat{p}_1(1 - \hat{p}_2)$, $(1 - \hat{p}_1)\hat{p}_2$, $\hat{p}_1\hat{p}_2$ and $1 - (1 - \hat{p}_1)(1 - \hat{p}_2)$. If we generate 999 such bootstrap samples, the 25th and 976th largest of them provide an approximate 95% confidence interval for N. In our example $\hat{N} = 125$ and $p_1 \equiv p_2 \equiv p$, with $\hat{p} = 0.523$.

For each resample, an estimate of the population size, $\hat{N}^*$, is computed using an appropriate estimator (Chapman's modified estimator in the case of our example).

Suppose that B samples are generated. Denote the ith largest estimate by $\hat{N}^*_{(i)}$. Then the bootstrap estimate of the 95% CI for N is given by $(\hat{N}^*_{(j)}, \hat{N}^*_{(k)})$, where $j = (B + 1) \times 0.025$ and $k = (B + 1) \times 0.975$.

The nonparametric bootstrap procedure is very similar in that we also compute B estimates using generated surveys from a population of size $\hat{N}$. The only change is that we resample the observed "capture histories" with replacement. A capture history is a row vector of 0s and 1s with a 0 in position s if the animal was not captured on occasion s, and a 1 in position s if it was captured. For our two-sample survey example, the $u_1 - m_2$ animals captured on the first occasion only have capture history $(1, 0)$, the u_2 captured on the second occasion only have the history $(0, 1)$, the m_2 captured on both occasions have the history $(1, 1)$, and the estimated $\hat{N} - (u_1 + u_2)$ animals that were never captured are assigned the capture history $(0, 0)$.

To generate a survey we take a simple random sample (with replacement) from the $\hat{N}$ animals and note their capture histories. From these we can compute n_1^*, u_2^* and m_2^* for the survey and hence $\hat{N}^*$. The bootstrap distribution and estimated CI of $(106, 151)$ from 999 bootstrap samples is shown in Figure 6.3. It is almost identical to the profile likelihood CI.

6.3.5 Heterogeneity

We know from Figure 3.7 that the example population is heterogeneous, but the likelihood equations developed so far, and the Petersen estimator, assume that on any one occasion all animals are equally catchable. We showed in Chapter 5 that neglecting heterogeneity when estimating abundance using removal methods could lead to substantial bias. It seems

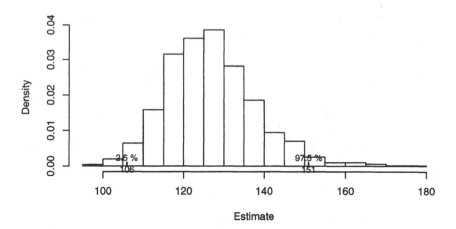

Figure 6.3. Bootstrap distribution of Chapman's modified estimator of N from a two–sample mark-recapture survey of the St Andrews example population. The 2.5th and 97.5th percentiles are indicated.

likely that neglected heterogeneity is causing the simple mark-recapture abundance estimate for the example population to be negatively biased.

To investigate this, we use the library WiSP again to simulate surveys and estimate abundance, and then draw conclusions about the properties of the estimator from the distribution of the simulated abundance estimates. For compatibility with the removal method simulations, we used exactly the same population and exactly the same capture function. (Recall that mark-recapture surveys contain within them removal surveys – with animals being removed from the unmarked population on each occasion – but also contain additional data, in the form of recaptures. Intuitively, we expect the additional data to improve the performance of the estimator in some way.)

Figure 6.4 shows the distribution of Chapman's modified two-sample mark-recapture estimator of N from 1,000 simulated surveys (a) when the population is heterogeneous and (b) when it is homogeneous. It is clear from the distributions that neglecting heterogeneity when it is present has led to substantial negative bias, and that the estimator is approximately unbiased when there is no heterogeneity. This is the same effect as was observed with the removal method survey simulations, and for the same reason. In Section 6.2.1 we noted that if marked and unmarked animals are equally catchable, then it is reasonable to use the proportion of marked animals that are recaptured on the second occasion as an estimator of the capture probability of unmarked animals. But if the population is heterogeneous, the more catchable animals will tend to be caught and marked first, and the capture probabilities of unmarked animals will tend to get smaller and smaller after each capture occasion – exactly the same phenomenon that

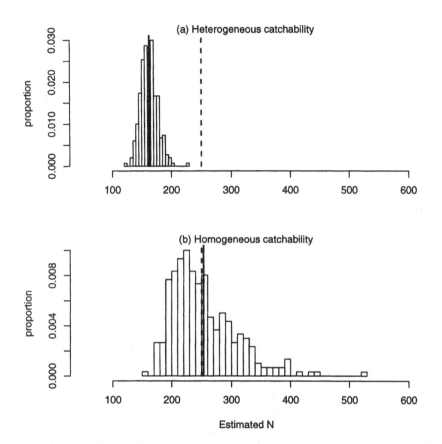

Figure 6.4. Simulated distribution of Chapman's modified two-sample mark-re-capture estimator of N. Plot (a) is the distribution when catchability is hetero-geneous, while (b) is when catchability is homogeneous (see text for details). The solid vertical line is at the mean of the simulated Ns; the dotted vertical line is at the true N.

caused bias with the removal method. Put another way, we estimate that we have captured a higher proportion of the unmarked animals than we really have, because the estimate is obtained from the marked animals, which tend to be more catchable.

Neglecting heterogeneity in mark-recapture surveys can result in very negatively biased abundance estimates. This is **the** difficulty with simple mark-recapture estimators of abundance, and the reason that in many applications they are not a useful method for estimating animal abundance. There are methods that accommodate heterogeneity: we consider some in Part III.

Is there evidence in the simulation results that the recapture data has improved the performance of the mark-recapture estimator compared with the simple removal estimator? Yes, although the methods seem equally biased (compare Figures 5.9 and 6.4). Firstly, none of the mark-recapture estimators are inadmissible, while some of the removal method estimates were. Secondly, the variance of the mark-recapture estimator is substantially smaller than the variance of the removal estimator. In many applications removal methods are used because animals are killed or otherwise physically removed from the population. In this case, using mark-recapture methods is not an option, but in some applications either method could be implemented. In these situations, mark-recapture methods are generally preferable.

6.4 Related methods and models

6.4.1 Unknown marking process

Note that if we knew nothing about the marking process on the first occasion, or if marking on this occasion occurred in some arbitrary fashion (rather than by catching every animal with probability p_1), then data from the first capture occasion tells us nothing about N except that it is at least the number of marked animals in the population by the start of the second occasion, $M_2(= u_1 = n_1)$. In this case we could estimate N from the likelihood Equation (6.7) alone, conditional on M_2.

6.4.2 Removal method likelihood

Note that $P_1 \times P_{2u}$ is identical to the two-sample removal method likelihood of Chapter 5. In retrospect this should not be surprising because you can think of capturing and marking an animal as removing it from the population of unmarked animals. So you can think of a mark-recapture experiment as containing a removal method experiment on unmarked animals within it. The removal method does not use the recapture information (contained in P_{2m}). So providing the assumptions of the methods hold, you would expect the mark-recapture method to perform better than the removal method because it contains additional data (i.e. m_2) in the likelihood component P_{2m}. The simulations above bear this out to the extent that the variance of the simple mark-recapture estimator is substantially lower than that of the simple removal method estimator. The estimated bias of the mark-recapture $\hat{N}$ differs very little from that of the simple removal method $\hat{N}$.

6.4.3   Hypergeometric models

In the models we developed above, we treated both the numbers of animals captured on each occasion (the n_ss) and the numbers of marked animals captured (the m_ss) as random variables. (Note that once n_s and m_s are known, u_s is also known.) That is, our likelihood functions give the likelihood of observing both the n_ss and the m_ss. A commonly-used alternative is to condition on the n_ss and use the conditional likelihood for the m_ss given the n_ss. That is, to work with the likelihood of observing the number of marked animals that are captured on each occasion, given the total sample size on the occasion. The latter leads to a simple hypergeometric model in the case of a two-sample mark-recapture experiment (see Exercise 6.5) and a product hypergeometric model in the multiple capture case.

The hypergeometric models are not as versatile as the likelihood developed above; they do not generalize easily to allow "trap-happiness" (increased probability of capture of animals that have previously been captured), or "trap-shyness" (decreased probability of capture of animals that have previously been captured), for example. The likelihoods developed above also correspond more closely with likelihoods developed for other methods in this book. We do not, therefore, pursue hypergeometric models further.

6.4.4   Single mark, multiple captures

In some applications, a single marking occasion followed by multiple recaptures is common. This is the case with band recovery methods for migratory birds, for example. A key aim of these methods is usually survival estimation in an open population context, but for closed populations they fit readily into the likelihood framework presented in this chapter. The only difference is that with one marking occasion, $M_s = M_1$ for all $s = 2, \ldots S$.

6.5   Multiple occasions: the "Schnabel census"

The MLE for two capture occasions ($S = 2$) has a simple form, which was given in Equation (6.1). There is an explicit MLE for the hypergeometric model in the case of a three capture occasions ($S = 3$). It is not quite as simple as the Petersen estimator, but we can still write it down (see Seber, 1982, p132). When there are more than three occasions ($S > 3$), we don't have an explicit MLE and we have to maximize the likelihood numerically. That is, we can't write down an equation for the MLE in general, but for any particular set of numbers of captures and recaptures from any particular experiment, we can find the MLE by searching the parameter space numerically until the location that the likelihood attains its maximum value is located. To do this, we do need an expression for the likelihood itself. We derive the general likelihood below.

6.5.1   A likelihood for multiple capture occasions

Technical
section
↓

We can extend the likelihood of Equation (6.8) to more than two occasions simply by taking the product over all occasions. In general, for S occasions:

$$L_S = \prod_{s=1}^{S} L_s = \prod_{s=1}^{S} P_{su} \times P_{sm} \tag{6.11}$$

$$= \prod_{s=1}^{S} \binom{U_s}{u_s} p_s^{u_s} [1 - p_s]^{U_s - u_s} \binom{M_s}{m_s} p_s^{m_s} [1 - p_s]^{M_s - m_s}$$

↑
Technical
section

6.5.2   Capture histories and individual identification

The likelihoods developed above are for the case in which we could observe whether or not a captured animal had been captured and marked before, but not on which occasion it was captured or its individual identity. This would be the case if animals were marked by painting a red dot on them, for example. On observing an animal with a red dot, we can tell that it has been captured before, but we can't say which animal it is, nor on which occasion(s) it was captured previously. The likelihoods for these two scenarios differ slightly from those derived above, but in a fairly straightforward way.

Capture histories

The marks might be such that we can tell from them the occasion(s) on which an animal was previously captured. This would be the case if we added a different coloured dot to caputured animals on each capture occasion, for example. In this case L_s in Equation (6.11) becomes

$$L_s = \prod_{\underline{\omega}_s} \binom{N_s(\underline{\omega}_s)}{n_s(\underline{\omega}_s)} p_s^{n_s(\underline{\omega}_s)} [1 - p_s]^{N_s(\underline{\omega}_s) - n_s(\underline{\omega}_s)} \tag{6.12}$$

where $\underline{\omega}_s$ is one of the 2^{s-1} possible capture histories of marked animals in the population by the start of occasion s and $\prod_{\underline{\omega}_s}$ is the product over all these capture histories. $N_s(\underline{\omega}_s)$ is the number of animals with capture history $\underline{\omega}_s$, and $n_s(\underline{\omega}_s)$ is the number of these that are caught.

Individual identities

Once caught, each individual might be recognizable on subsequent occasions. This would be the case if every captured animal had a unique number attached/painted onto it, for example. In this case L_s in Equation (6.11) becomes

$$L_s = \prod_{i=1}^{M_s} p_s^{\omega_{si}}[1 - p_s]^{1-\omega_{si}} \tag{6.13}$$

where ω_{si} is a capture indicator variable which is equal to 1 if animal i was captured on occasion s and is zero otherwise.

6.6   Types of mark-recapture model

6.6.1   Classification by observation model

Pollock (1974) and Otis et al. (1978) developed a useful classification scheme for mark-recapture models for closed populations. It accommodates three basic sources of heterogeneity in capture probabilities. They are referred to as model "M_m", where m is a letter indicating the source of heterogeneity. Including a model with no heterogeneity, the basic models are as follows:

M_0 The population is assumed to be homogeneous with respect to capture probability, i.e. all animals are always equally catchable (subscript 0 for zero heterogeneity).

M_t Capture probabilities vary from one capture occasion to another, but all animals are assumed to be equally catchable on any one occasion (subscript t for heterogeneity depending only on time). This is the model for which we developed a likelihood function above.

M_b The only thing affecting capture probability is a response to capture; animals become "trap happy" or "trap shy" (subscript b for behavioural response).

M_h This allows individual animal heterogeneity only, i.e. animals are assumed to have different capture probabilities, but these remain constant over all capture occasions (subscript h for heterogeneity).

Four more complicated models are constructed by combining the three sources of heterogeneity in various ways, i.e. M_{th}, M_{tb}, M_{bh} and M_{tbh}. These are all models only in a rather vague sense; each accommodates a host of possible models with the given source(s) of heterogeneity in capture probability. But the classification is a useful way of structuring a discussion of mark-recapture models for closed populations.

Notice that the likelihood functions we derived above are based purely on observation models. Notice also that the model classification above is based only on observation models; it is purely on the basis of what factors affect detection/capture probability. A large part of the reason for this is no doubt that most closed-population mark-recapture analyses treat N as a fixed parameter and assume that the whole survey region is surveyed. In this case all the uncertainty about the number of animals in the survey region comes from the observation model. Given an observation model, there is no need for a state model when the whole region is surveyed (unless there is a specific interest in the processes determining the states), and model classification based on the capture model alone is adequate.

There is, however, a problem. For any model with an h in its subscript, the observation model, and hence the likelihood, can't be evaluated without catching the whole population. The observation model usually depends on some explanatory variables and includes unknown parameters. (With the M_t model and likelihood function we developed above, for example, the only explanatory variable is survey/capture occasion, s.) The difficulty with h-models is that the explanatory variables occur at the level of the animals (every animal can have a different explanatory variable), and until we observe every animal, we don't know the associated explanatory variable(s) and so can't evaluate the observation model. (Equation (B.6) in Appendix B is an observation model for h-models.) If we can't evaluate the observation model, we can't evaluate the associated likelihood function, so fully likelihood-based inference is impossible. We are left with two options: include a state model for the explanatory variable(s), or use design-based inference or other methods which do not involve the likelihood, to estimate N.

Most mark-recapture methods for h-type models do not use model-based inference for N (although estimation of p is model-based). We do not consider h-type models in this chapter. In Chapter 11, we consider both design-based and model-based inference for them. Model-based inference involves incorporating a state model into the likelihood function, and this leads to some interesting generalizations.

There is one special sort of animal-level variable for which we can evaluate the observation model, and this is capture history. Unlike other animal-level variables, we know the capture history of all unobserved animals in the population. This is a fundamental difference and it is therefore useful to distinguish capture history from other animal-level variables. In this chapter, we consider models which involve capture probabilities that depend on capture histories; consideration of models in which other sorts of animal-level variables affect capture probability is postponed to Chapter 11.

6.6.2   Models M_0, M_t, M_b, M_{tb}

For model M_0 we assume that capture probability is the same for all animals and all capture occasions.[1] For model M_t we assume that capture probability is the same for all animals but changes from one capture occasion to another. For model M_b we assume that capture probability depends only on whether the animal has previously been captured. Finally, for model M_{tb} we assume that capture probability is different for different capture occasions and depends on whether the animal has previously been captured.

6.6.3   Likelihoods for models M_0, M_t, M_b, M_{tb}

Capture functions

Technical section
↓

The only difference between these four models (and between all the classes of model described above) is how the detection or capture function, p is parametrized. Remember from Chapter 3 that the general form for the capture function for animal i on occasion s allows capture probability to depend on both the animal-level variables and the survey-level variables, $\underline{l}_s$, which apply on occasion s. Here the only animal-level variable we consider is capture history, $\underline{\omega}_{si}$ for animal i on occasion s. The most general capture function we consider in this chapter for animal i on occasion s can therefore be written as $p(\underline{\omega}_{si}, \underline{l}_s)$. The likelihood functions developed above still apply, with one change; p_s is replaced by $p(\underline{\omega}_s, \underline{l}_s)$ if capture history is observable, and by $p(c_s, \underline{l}_s)$ if it is not (where $c_s = 1$ if the animal has been marked by the start of occasion s, and $c_s = 0$ otherwise).

M_0: $p(\underline{\omega}_{si}, \underline{l}_s) = \theta$, and θ is the only capture function parameter to be estimated.

M_t: $p(\underline{\omega}_{si}, \underline{l}_s) = p(\underline{l}_s)$. One obvious form for $p(\underline{l}_s)$ has $\underline{l}_s = \theta_s$, and then $\underline{\theta} = (\theta_1, \dots, \theta_S)'$ is the vector of capture function parameters to be estimated. Another form, with fewer parameters, might have $\underline{l}_s$ equal to a scalar measure of catch effort (the number of traps used, for example). In this case, one

---

[1] Seber (1982, p 164) and others have pointed out that nothing is gained by using model M_0 instead of model M_t; nevertheless, we consider model M_0 here for illustration.

reasonable form for the capture function might be $p(l_s) = 1 - e^{-\theta l_s}$.

This form of capture function is commonly used in fisheries models (see Chapter 5). Many other forms of dependence of the detection probability on l_s are possible. Note that (i) the animal index, i, is redundant because capture probability is not a function of the animal-level variables, and (ii) in the derivation of the likelihood functions for this model, we used p_s in place of $p(l_s)$, for brevity.

M_b: $p(\underline{w}_{si}, l_s) = p(\underline{w}_{si}) = p(c_{si})$: $c_{si} = 1$ if by the start of occasion s animal i has been marked, and $c_{si} = 0$ if not. An obvious form of dependence of the capture probability on $\underline{w}_{si}$ is $p(\underline{w}_{si}) = \theta_{c_{si}}$, in which case $\underline{\theta} = (\theta_0, \theta_1)'$ is the parameter vector to be estimated. Other forms may be appropriate, depending on the application.

M_{tb}: $p(\underline{w}_{si}, l_s) = p(c_{si}, l_s)$ depends on both the animal-level variable (c_{si}) and a survey-level variable $(l_s = \theta_s)$. One useful form for dependence of the capture probability on these two variables is the logistic:

$$p(\underline{w}_{si}, l_s) = \frac{e^{(\theta_c \times c_{si} + \theta_s)}}{1 + e^{(\theta_c \times c_{si} + \theta_s)}} \qquad (6.14)$$

where θ_c is a behaviour parameter for captured animals, and $c_{si} = 1$ if animal i has been captured by the start of occasion s, and $c_{si} = 0$ otherwise. Many other forms are possible.

↑
Technic[
section]

Note that you can't estimate both behaviour and time effects in model M_{bt} without additional assumptions. You can see this by thinking about how many bits of data you have, and how many parameters you need to estimate.

With a two-sample survey you have three bits of data (n_1, u_2, m_2) and four parameters to estimate: N, capture probabilities p_1, p_2 for unmarked animals on the two occasions, and capture probability p_{c_2} (say) for marked animals on the second occasion (there are no marked animals on the first occasion). You can't sensibly estimate four parameters with only three bits of data.

With a three-sample survey you have five bits of data $(n_1, u_2, m_2, u_3, m_3)$ but now you have six parameters to estimate: $N, p_1, p_2, p_3, p_{c_2}, p_{c_3}$. You can't sensibly estimate six parameters with only five bits of data. And so on.

Equation (6.14) contains an assumption that reduces the number of parameters to an estimable number when there are three or more capture occasions (from which you get five or more bits of data). The assumption

Table 6.1. St Andrews dataset: numbers of marked and unmarked animals detected on each occasion (i.e. by each observer).

	Occasion/observer number							
	1	2	3	4	5	6	7	8
Number unmarked detections	64	33	18	9	11	4	5	18
Number marked detections	0	34	55	50	58	54	58	75
Total	64	67	73	59	69	58	63	93

is that capture changes catchability in the same way for all capture occasions – by the single term θ_c in the equation. In this case, the parameters for the three-capture occasion are N, θ_c, $\theta_{s=1}$, $\theta_{s=2}$ and $\theta_{s=3}$, so we no longer have more parameters than bits of data [2]. However, the assumption that recapture probabilities bear a constant relationship to initial capture probabilites over all capture occasions, for all animals, is highly questionable in most applications.

Likelihood functions

Once you have specified a capture function, the likelihood for these models is just the appropriate general likelihood function given earlier in this chapter, but with the appropriate capture function in place of p_s.

6.7 Examples

6.7.1 St Andrews example data revisited

We are able to identify recaptures in the St Andrews example dataset by their location: because the animals in this population are completely stationary, if two observers saw animals at the same location, they must be the same animals. Table 6.1 summarizes the mark-recapture data for the eight capture occasions.

Using the R library WiSP, we fitted models M_0, M_t and M_b to these data. The maximum log-likelihood values, AIC, and estimated abundance $\hat{N}$ from each of these models is given in Table 6.2.

Model M_b fits the data best – it has the largest log-likelihood of the three models. It also has the lowest AIC. On this basis, it is the best model of the three.

The nonparametric bootstrap distribution of abundance for model M_b is shown in Figure 6.5, together with the estimated 95% CI (164; 190). It is a

---

[2] Aside: when you can't estimate all the parameters from the data, the parameters are said to be not "identifiable".

Table 6.2. St Andrews dataset: Log-likelihood, AIC, and estimated abundance for models M_0, M_t, and M_b.

| | Model | | |
Statistic	M_0	M_t	M_b
log-likelihood	−75.1	−64.2	−59.2
AIC	154.2	146.4	124.4
$\hat{N}$	164	164	173

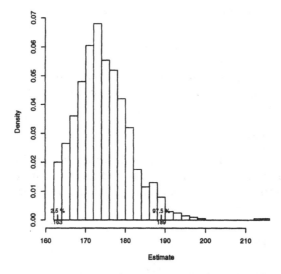

Figure 6.5. Nonparametric bootstrap distribution of the M_b abundance estimate for the St Andrews example dataset. The percentile method 95% confidence limits are marked on the bottom axis.

little worrying that the CI is so far from 250 (the true abundance). Given what we know about the St Andrews example population, however, this is not very surprising. We know that the population was heterogeneous: animals have different exposures, different sexes, and occur in different group sizes. Each of these factors might well affect how detectable (catchable in mark-recapture terms) they are. None of the models used above accommodates animal-level heterogeneity, and all of these models are negatively biased in the presence of such heterogeneity.

The model M_b has managed to accommodate the heterogeneity to some extent. It does so by way of a higher estimated capture probability for marked animals ($\hat{p}_m = 0.47$) than for unmarked animals ($\hat{p}_u = 0.28$). The more detectable animals tend to be caught on the early capture occasions, so that they are over-represented in the marked population. Marked ani-

Table 6.3. Simulated survey of homogeneous population with $N = 1,000$.

	Occasion/observer number							
	1	2	3	4	5	6	7	8
Number unmarked detections	91	62	53	39	30	19	13	10
Number marked detections	0	7	7	10	7	2	7	2
Total	91	69	60	49	37	21	20	12

Table 6.4. Simulated survey of homogeneous population with $N = 1,000$: log-likelihood, AIC, and estimated abundance for models M_0, M_t and M_b.

		Model	
Statistic	M_0	M_t	M_b
log-likelihood	−98.8	−35.9	−44.5
AIC	201.6	89.7	94.9
$\hat{N}$	1,246	1,192	346

mals are, as a result, more catchable than unmarked animals on average, although this is a consequence of heterogeneity rather than trap-happiness.

In this example there is only a 5% difference in the abundance estimates for the three models. In some cases the difference between estimates for different models can be large and in this case model selection is all the more important. In cases where there remains substantial uncertainty about which is the most appropriate model, and choice of model makes a large difference to the abundance estimate, it is wise to incorporate the model uncertainty in the estimated confidence interval. This can be done in a number of ways. The simplest is to include choice of model in the nonparametric bootstrap loop. That is, for each bootstrap resample, fit all the models under consideration, evaluate each model's AIC, and use the estimate from the model with lowest AIC. See Buckland *et al.* (1997) for discussion of this and related issues.

6.7.2   No animal heterogeneity; true model M_t

Using the R library WiSP, we generated a population of 1,000 animals with no heterogeneity, surveyed it with eight capture occasions using model M_t, and fitted models M_0, M_t and M_b to the survey data. Table 6.3 summarizes the survey data and Table 6.4 shows the maximum log-likelihood values, AIC, and estimated abundance $\hat{N}$ from each of the models .

Model M_t is the true model and model M_t fits the data best (it has the largest log-likelihood) and has the lowest AIC. The true abundance ($N =$

Table 6.5. Simulated survey of homogeneous population with $N = 200$.

	Occasion/observer number							
	1	2	3	4	5	6	7	8
Number unmarked detections	18	10	12	11	5	6	3	3
Number marked detections	0	3	0	2	2	0	3	1
Total	18	13	12	13	7	6	6	4

Table 6.6. Simulated survey of homogeneous population with $N = 200$: Log-likelihood, AIC, and estimated abundance for models M_0, M_t and M_b.

	Model		
Statistic	M_0	M_t	M_b
log-likelihood	-35.9	-27.3	-28.4
AIC	75.9	72.5	62.7
$\hat{N}$	224	218	77

1,000) is well within the nonparametric bootstrap confidence interval of (919, 1,608). Note that an erroneous choice of model M_b would result in an abundance estimate that is only a third of the true abundance!

6.7.3   No heterogeneity; true model M_t; small n

For this example, we generated a population like that in the example above, but containing only 200 animals. We again surveyed it with eight capture occasions, using an identical observation model to that used above, and again fitted models M_0, M_t and M_b. Table 6.5 summarizes the survey data and Table 6.6 shows the maximum log-likelihood values, AIC, and estimated abundance $\hat{N}$ from each of the models.

The true model is M_t and as might be expected, M_t fits the data best (it has the largest log-likelihood). But model M_b has the lowest AIC, by quite a large amount, and it gives an abundance estimate that is only a third of those from M_0 and M_t – both of which are close to the true N of 200.

Why has the AIC led to the wrong model? It is not guaranteed to give the correct model (which we never know in reality), but in this case the choice of an incorrect model is due mostly to small sample size. Recall that AIC is composed of a goodness-of-fit term $(-2\ln(L))$ and a penalty for having to estimate many parameters $(+2q)$. Model M_t has the best fit, but with eight capture occasions, it involves estimation of eight capture probability parameters (one for each occasion), whereas model M_b involves estimation of only two capture probability parameters (one for unmarked animals and one for marked animals). The cost of estimating an additional six parameters with M_t is judged by the AIC criterion not to warrant the improvement in fit over M_b that is obtained by doing so. A larger sample size gives parameter estimates that have higher precision, and more parameters can be estimated before AIC judges the loss in precision due to estimation of more parameters not to be worthwhile.

6.8   Effect of violating main assumptions

Assumption 1 Capture probability is

M_0 constant.

M_t depends only on capture occasion.

M_b is constant except that marked animals are more/less catchable than unmarked.

M_{tb} depends on capture occasion and whether or not an animal has been caught.

Effect of violation: All estimators are biased by violation of the assumption above on which they are based. A particular problem is that catchability is very often different for different animals (there is unmodelled heterogeneity in the population) and this can make the estimators above very biased. Even if heterogeneity is not a problem, estimators can be substantially biased (as in the small sample simulated survey considered above). It is important to try a variety of plausible models; AIC can be used to select between them.

Assumption 2 Animals are captured independently of one another.

Effect of violation: CIs based on the assumption tend to be biased (usually too narrow) and point estimates may be biased. Violation of this assumption tends to be a problem if practicalities make it so. For example, if a trap can hold only one animal, then when all traps are full, the probability of catching any of the uncaptured animals is zero; capture probability depends on how many animals have been captured. Care should be taken in designing the experiment to avoid this sort of problem.

Assumption 3 No marks are lost.

Effect of violation: Mark loss results in positive bias in abundance estimation. The lower recapture frequency that mark loss causes is interpreted by the estimators as if a smaller fraction of the population had been marked than was actually the case. The same effect occurs if marks are not reported for some captured animals even though they are still on the animals.

6.9   Summary

Like the removal methods of Chapter 5, the mark-recapture methods of this chapter do not involve state models. The whole survey region is assumed to be covered, and no differences between animals (aside from their capture histories) are recognized. The differences between the various mark-recapture models of this chapter are just differences in the way capture probability is modelled.

While the estimators perform better than removal method estimators because they use information from recaptures as well as "removals" from the unmarked population, like removal method estimators, they are subject to potentially large bias from unmodelled heterogeneity. If the captured animals tend to be the more catchable animals (as is often the case), the catchability of uncaptured animals will be overestimated, and abundance will be underestimated. In Chapter 11, we deal with mark-recapture methods that model heterogeneity (animal-level differences in capture probability).

In the next chapter we deal with the only (relatively) simple methods that are robust to heterogeneity, namely distance sampling methods.

6.10   Exercises

Exercise 6.1   By considering "marked" and "unmarked" to be animal type, show that the two-sample change-in-ratio estimator of N given in Equation (5.12) is identical to the Petersen mark-recapture estimator given in Equation (6.1) of this chapter.

Exercise 6.2   Consider a two-sample mark-recapture experiment in which capture probabilities are the same on both capture occasions.

(a) Using the approximation $\frac{d\ln(N!)}{dN} = \ln(N)$, show that the MLEs for N and p obtained from the two-sample mark-recapture likelihood Equation (6.8) with $p_1 \equiv p_2 \equiv p$ are solutions to the equations

$$N = \frac{n}{p(2-p)} \quad \text{and} \quad p = \frac{n_1 + n_2}{2N} \qquad (6.15)$$

where n_1 animals are captured on occasion 1, n_2 are captured on occasion 2, and n different animals are captured in all.

Note that the second of these equations specifies the p that maximizes the likelihood at any given N. Substituting $\frac{n_1+n_2}{2N}$ for p in the likelihood equation therefore gives the profile likelihood for N.

(b) Using Equations (6.15), show that the MLEs for N and p are

$$\hat{N} = \frac{(n_1 + n_2)^2}{4m_2} \qquad \text{and} \qquad \hat{p} = \frac{m_2}{\frac{(n_1+n_2)}{2}} \qquad (6.16)$$

where m_2 is the number of marked animals that are recaptured on the second occasion.

Exercise 6.3   Consider a two-sample mark-recapture survey with different capture probabilities on each occasion (p_1 on occasion 1, p_2 on occasion 2). Show that

(a) $\Pr\{u_1, u_2, m_2 | N, p_1, p_2\}$ is multinomial.

(b) You can estimate N without estimating p_1; so you can estimate N even when you have no idea how marks came to be on the marked animals.

Exercise 6.4   Consider a multinomial distribution for the outcomes of N independent events with four possible outcomes and cell (outcome) probabilities θ_1, θ_2, θ_3 in the first three cells, and $1 - \sum_{i=1}^{3} \theta_i$.

(a) Given counts $n(10)$, $n(01)$ and $n(11)$ in the first three cells, maximize the log of the corresponding multinomial likelihood with respect to N and $\underline{\theta} = (\theta_1, \theta_2, \theta_3)$, using the approximation $\frac{d \ln(N!)}{dN} = \ln(N)$, to show that the MLEs for N and $\underline{\theta}$ satisfy

$$\hat{N} = \frac{n}{p.} \qquad \text{and} \qquad \hat{\underline{\theta}} = \left(\frac{n(10)}{\hat{N}}, \frac{n(01)}{\hat{N}}, \frac{n(11)}{\hat{N}} \right) \qquad (6.17)$$

where $n = n(10) + n(01) + n(11)$ is the sum of counts in the first three cells, and $p. = \sum_{i=1}^{3} \theta_i$.

(b) Can N and $\underline{\theta}$ be estimated from Equations (6.17)?

(c) Show that the two-sample mark-recapture likelihood Equation (6.8) can be written as a multinomial likelihood with cell probabilities $\theta_1 = p_1(1 - p_2)$, $\theta_2 = (1 - p_1)p_2$ and $\theta_3 = p_1 p_2$, and abundance parameter N.

(d) Using the relationship between the θs and the capture probabilities, show that the Petersen estimator is the MLE for N.

Exercise 6.5

(a) A jar contains B black balls and W white balls. A sample of n balls is taken from the jar, without replacement, in such a way that every ball is equally likely to be chosen. Write down the pdf of b, the number of black balls in the sample.

(b) If black balls represent marked animals, and white balls represent unmarked animals, what are B, W, n and b in the mark-recapture notation used in this chapter?

(c) Using the notation from (b), write down the conditional distribution of the number of recaptures in a two-sample mark-recapture survey, given that the number of marked animals in the population by the start of the second capture occasion, and the total number of animals that were captured on the second occasion.

Exercise 6.6   Consider a mark-recapture survey with S sampling occasions, in which animals are marked in a way that allows the capture history of every animal ever captured to be observed. Show that the likelihood Equation (6.12) is multinomial, with 2^S cells, one for each possible capture history.

7
Distance sampling

Key idea: estimate p by modelling decline in detection frequency with distance.

Number of Surveys:	Only one survey required.
State Model:	Animals are distributed independently of line or point location.
Observation Models:	Detection probability decreases with distance; all animals on the line or point are detected; independent detections.

7.1 Introduction

With distance sampling we can estimate abundance from a single survey because the method involves a strong (but often reasonable) assumption about detection probability. Distance sampling comprises several related methods which involve measuring or estimating distances of detected animals from a line or point. The two main methods are line transect sampling

---

Likelihood function and key notation:

$$L(N;\underline{\theta}) = \binom{N}{n}(1 - \pi_c E[p])^{N-n}(\pi_c)^n \prod_{i=1}^{n}\pi(x_i)p(x_i)$$

N: animal abundance.

n: number of animals detected.

π_c: $= \frac{a}{A}$: probability that an animal is in the covered region.

x: distance of animal from line or point.

$\pi(x)$:   pdf of x for all animals in the covered region
$= 1/w$ for line transect sampling
$= 2x/w^2$ for point transect sampling.

$p(x)$:   probability of detecting an animal at distance x.

$\underline{\theta}$:   Parameters of the detection function, $p(x)$.

$E[p]$:   expectation over x (with respect to the state model) of $p(x)$ in the covered region.

w: maximum distance at which animals are recorded.

---

and point transect sampling (sometimes called variable circular plots). Both are extensions of plot sampling, in which only incomplete counts of animals within the covered region are made. The key assumption about detection probability is that all animals that are located on the line or at the point are certain to be detected. Detection probability $p(x)$ is assumed to fall off in a smooth way out to some distance $x = w$ from the line or point.

We consider each of line transect and point transect sampling in detail below. We then review the related techniques of cue counting, trapping webs and indirect distance sampling surveys.

7.2   Line transect sampling

Line transect sampling is an extension of strip sampling. Suppose there are J strips, each of width $2w$ (so that the strip extends w either side of the centreline). Suppose further that strip j has length l_j, with $\sum_{j=1}^{J}l_j = L$. Then the covered region is of size $a = 2wL$. In the case of strip sampling, if we count n animals in these strips, then we estimate population size N by $\hat{N} = n/(a/A)$ (Chapter 4). In line transect sampling however, the number of animals detected is not the total number present in the covered region, so there is another layer of complexity to the modelling problem.

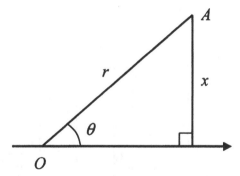

Figure 7.1. The observer, currently at O, detects an animal at A. If x is difficult to measure directly, the observer may record the radial distance r and sighting angle θ, from which $x = r\sin(\theta)$.

In line transect sampling, the observer travels along the centreline of the strip, and records the perpendicular distance x of each detected animal (or animal group) from the line. Sometimes, it is easier to record the detection radial distance r and angle θ (Figure 7.1), from which $x = r\sin(\theta)$. The design generally comprises lines, or a systematic grid of lines, randomly placed in the survey region. The random placement ensures that animals are uniformly distributed with respect to their distances from the line. It is assumed that all animals on or very close to the line are detected, but probability of detection may decrease with distance from the line, out to some distance w.

7.2.1   A simple line transect survey

We consider a line transect survey of the St Andrews example population, with four transect lines, running down the centres of the same four strips that were used in the plot survey example of Chapter 4. The lines and the locations of the 88 detected animal groups are shown in Figure 7.2.

Observers searched 4 m either side of each line, so that the covered region consists of the four strips of width 8 m, with the lines running down their centre. Exactly one third of the $N_c = 132$ animals in the covered region were missed and by comparing Figures 7.3 and 7.4 it is clear that more groups were missed far from the line than close to it. Not surprisingly, detection probability decreases as distance from the observer increases. By modelling the decrease, line transect methods are able to give estimates of the proportion of groups missed.

The line transect estimate of group abundance in the covered region ($\hat{N}_c = 118$) is 11% lower than the true number of groups in the covered region ($N_c = 132$). True abundance is well within the estimated 95% confidence interval of (81; 173). (We expand below on how confidence intervals

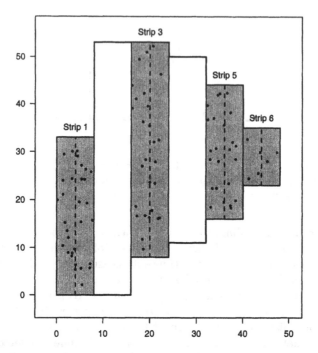

Figure 7.2. St Andrews example data, location of transects (dashed lines) and detected animals.

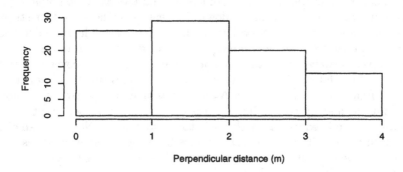

Figure 7.3. St Andrews example data, perpendicular distance distribution of the $n = 88$ animals detected in the covered region.

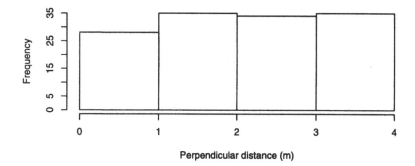

Figure 7.4. St Andrews example data, perpendicular distance distribution of the $N_c = 132$ animals in the covered region.

are obtained.) This example is not a good example of real line transect surveys because it involves so few lines and covers such a large fraction of the survey area. With just four sampled strips, the confidence interval is rather wide; the lower limit is less than the number of groups detected! Nevertheless, the fact that the population is heterogeneous (some animals are much more detectable than others) does not seem to have caused substantial negative bias in estimating N, as it did with the removal and mark-recapture method examples.

The estimated detection function is shown in Figure 7.5. The histogram suggests how detection probability falls off with distance, but this representation of the detection function is "noisy" because the heights of the bars vary from the underlying detection function by chance. The underlying "signal", i.e. shape of the detection function, is obscured somewhat by this "noise". The smooth fitted curve is our estimate of the underlying shape of the detection function. (Details of how it was obtained are given below.)

If all animals on the transect line are detected, a reliable estimate of the detection function gives a reliable estimate of the proportion of animals in the covered region that were missed (Figure 7.6).

Providing the main assumptions of the method hold, line transect methods are able to "compensate" for missed animals in the covered region and give unbiased estimates of abundance even when a large proportion of the animals in the covered region are missed.

In the case of plot sampling, we obtained an estimate of abundance in the survey region by dividing the observed count in the covered region (N_c) by the probability that a group was in the covered region (a/A). We get an estimate of abundance in the same way here, but using the estimated abundance in the covered region instead of the true abundance in the covered region (which was not observed): $\hat{N} = \hat{N}_c/(a/A) = 118/0.562 =$

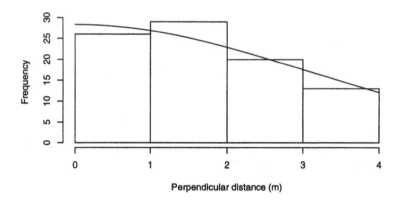

Figure 7.5. St Andrews example data, histogram of detection distances and corresponding estimated detection function (scaled to frequencies).

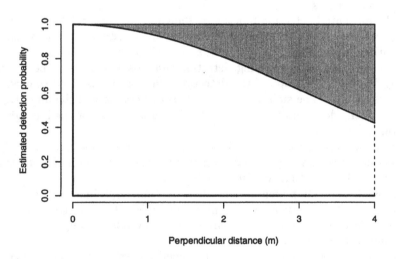

Figure 7.6. St Andrews example data, estimated detection function. The shaded
area corresponds to animal groups that we estimate are undetected.

215. The true abundance falls well within the estimated confidence interval of (144; 308).

There are two distinct sources of randomness operating: that associated with the observation model and that associated with the state model. Estimated abundance within the covered region depends crucially on the observation model, and a key part of this model is the detection function. Much of the history of line transect methodological development concerned the development of reliable, flexible methods for estimating the detection function.

Given estimated abundance within the covered region, inferences about abundance outside the covered region depend on the survey design and, in the case of model-based inference, the state model. In any event, it is important that the design gives a covered region that is representative of the survey region in some sense; usually the sense is that any point in the survey region is as likely to have been covered as any other point.

7.2.2   Maximum likelihood estimation

Distance sampling methods involve both state models and observation models. Plot sampling involves a state model because the covered region is smaller than the survey region, but it does not involve an observation model because by assumption all animals in the covered region are detected. Simple removal and mark-recapture methods involve observation models because not all animals in the covered region are detected, but they do not involve state models, because by assumption the whole survey region is covered and individual animal characteristics (which are determined by a state model) do not affect detection probability. In distance sampling, not all animals in the covered region are detected, so we need an observation model to draw inferences about abundance within the covered region. The covered region is typically much smaller than the survey region and there is substantial uncertainty about animal characteristics affecting capture probability because detection probability depends on distance from the line, and the locations of all animals relative to the line are not known.

To set line transect sampling into a full likelihood framework, we need to specify a state model for the locations of all groups at the time of the survey, and an observation model for the detection process within the covered region.

State and observation models

As with plot sampling, if a group is equally likely to be anywhere in the survey region of size A, then the probability that it falls in a covered region of area a is a/A. In this case, the covered region is a strip (or series of strips) of half-width w and total length L, so that $a = 2wL$. For the simple example above, $a/A = 944/1640$. The probability density of the perpendicular distance from the transect line (x) of a group in the strip (i.e. within w of the line) is uniform. That is,

$$\pi(x) \;\; = \;\; 1/w \qquad (7.1)$$

For the simple example above, $w = 4$ so that $\pi(x) = 1/4$.

The probability that a group is in the strip and in the perpendicular distance interval $(x, x + dx)$ (where dx is suitably small) is therefore $a/A \times \frac{1}{w} dx$. This is our state model for a single group.

The probability that a group is detected, given that it is at x, is $p(x)$. This is the basis of our observation model, and we use it with the state model to derive a full likelihood below.

Encounter rate component of the likelihood

Suppose we knew that the detection probability of animals within the covered region was some value, which we will call $E[p]$. The probability that an animal is in the covered region is $\pi_c = a/A$, so the probability that an animal is in the covered region **and** is detected, is $\pi_c \times E[p]$. Now if groups are detected independently of one another, then the probability that n groups are detected is a binomial random variable with parameters N and $\pi_c E[p]$. So if we knew $E[p]$ the likelihood for N would be

$$L(N|E[p]) \;\; = \;\; \binom{N}{n} (\pi_c E[p])^n \left[1 - \pi_c E[p]\right]^{N-n} \qquad (7.2)$$

Given $E[p]$, the MLE of N is $n/(\pi_c E[p])$. This likelihood (and MLE) is identical to that of Chapter 4 when $E[p] = 1$. However, in reality we don't actually know $E[p]$, and the likelihood has two unknown parameters (N and $E[p]$) with only one bit of data, n. We need further assumptions and/or further data to estimate N sensibly. The key to estimating N from line transect data is estimating $E[p]$, and it is the additional data provided by the perpendicular distances to detections that allow us to do this. To use the data contained in the observed perpendicular distances, we need a statistical model for how the observed distances were generated. We derive this below, but first we express $E[p]$ in terms of the intuitively appealing "effective strip width", or more correctly, the "effective strip half-width" which is often abbreviated to "esw".

Effective strip half-width

The "effective strip half-width" (which we denote μ) is the effective half-width of the strip, in the sense that the expected number of animals detected at distances greater than μ from the line equals the expected number missed within μ of the line (Figure 7.7). The effective area covered is $2\mu L$.

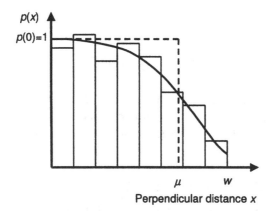

Figure 7.7. The effective strip half-width μ is the distance for which as many animals are detected beyond μ as are missed within μ. Hence the area of the rectangle, $p(0) \times \mu$, which equals μ when detection on the line is certain, is equal to the area under the curve, $\int_0^w p(x)\,dx$.

In other words, if you did a plot survey, detecting everything within μ of the line, you would expect to see the same number of animals as if you did a line transect survey with esw equal to μ. You effectively cover an area of $2\mu L$ in both cases.

$E[p]$ is the proportion of the covered region that you effectively survey, i.e. the effective covered area divided by the covered area:

$$E[p] \;=\; \frac{2\mu L}{2wL} \;=\; \frac{\mu}{w} \qquad (7.3)$$

We give a more mathematical rationale below.

Technical section
↓

The probability that a group is in the strip, is in the perpendicular distance interval $(x, x + dx)$, and is detected, is $p(x) \times a/A \times \pi(x)\,dx$. The probability of detecting a group, given that it is in the strip at an unknown perpendicular distance, is the average detection probability of animals within the strip; that is, it is the expected value of the detection function in the strip with respect to the state model:

$$E[p] \;=\; \int_0^w p(x)\pi(x)\,dx \qquad (7.4)$$

In the case of line transect surveys, $\pi(x) = \frac{1}{w}$ and

$$E[p] \;=\; \int_0^w p(x)\frac{1}{w}\,dx \;=\; \frac{1}{w}\mu \qquad (7.5)$$

where

$$\mu \;=\; \int_0^w p(x)\,dx \qquad (7.6)$$

and $p(x)$ is the probability that a group, distance x from the line $(0 \le x \le w)$, is detected. Both $E[p]$ and $p(x)$ depend on an unknown parameter vector $\underline{\theta}$, which for brevity we do not show explicitly.

↑
Technical
section

Perpendicular distance component of the likelihood

From the development above, we have a likelihood and a MLE for N if only we knew the parameter(s) $\underline{\theta}$ of the detection function. Clearly the perpendicular distance data contain information about the detection function: if the histogram of detected distances drops off fast, average detection probability is low; if it stays high, average detection probability is high. We now need a likelihood involving the distances, that will allow us to estimate the detection function (or, equivalently, $\underline{\theta}$). We refer to this as the "conditional likelihood" because it is conditional on the number of animals detected, n; the randomness in n is captured in the encounter rate likelihood of Equation (7.2). The conditional likelihood is derived below.

Technical
section
↓

The probability that a group in the strip is detected, given that it is at x, is $p(x)$. The probability density of x for groups in the strip is $\pi(x)$. Hence

$$
\begin{aligned}
Pr\{x|\text{seen}\} \;&=\; \frac{Pr\{\text{seen}|x\}\pi(x)}{\int_0^w Pr\{\text{seen}|x\}\pi(x)\,dx} \\[2mm]
&=\; \frac{p(x)\pi(x)}{\int_0^w p(x)\pi(x)\,dx} \;=\; \frac{p(x)\pi(x)}{E[p]} \qquad (7.7)
\end{aligned}
$$

This is the probability density of x, given that a group has been detected, and we denote it $f(x)$. Note that $f(x) = \frac{p(x)}{\mu}$ when $\pi(x) = \frac{1}{w}$. In this case, $f(x)$ is simply the detection function rescaled so that it integrates to one: $\int_0^w \frac{p(x)}{\mu}\,dx = 1$.

If groups are detected independently, the probability density of xs of detected groups is $\prod_i f(x_i)$ and the likelihood for $\underline{\theta}$, given the xs of detected animals, is therefore

$$L(\underline{\theta}) \;=\; \prod_{i=1}^n f(x_i) \;=\; \left(\frac{1}{E[p]}\right)^n \prod_{i=1}^n p(x_i)\pi(x_i) \qquad (7.8)$$

When $\pi(x) = \frac{1}{w}$,

$$L(\underline{\theta}) = (\mu)^{-n} \prod_{i=1}^{n} p(x_i) \qquad (7.9)$$

↑
Technical
section

To recap: we can estimate $\underline{\theta}$, and hence $E[p]$, from the conditional likelihood Equation (7.8). We can then use this estimate of $E[p]$ in the encounter rate likelihood Equation (7.2) to estimate N. Alternatively, we could combine Equations (7.8) and (7.2) to get a full likelihood, and estimate N and $\underline{\theta}$ simultaneously from this. We develop both approaches in a bit more detail below.

MLE from the full likelihood

The likelihood for N and $\underline{\theta}$ is the product of Equations (7.2) and (7.8):

$$L(N;\underline{\theta}) = L(N|\underline{\theta}) \times L(\underline{\theta}) \qquad (7.10)$$

$$= \binom{N}{n}(1 - \pi_c E[p])^{N-n}(\pi_c)^n \prod_{i=1}^{n} p(x_i)\pi(x_i)$$

This is a full likelihood for distance sampling methods in general. When $\pi(x) = \frac{1}{w}$, it is a full likelihood for line transect methods.

To maximize the likelihood with respect to N and $\underline{\theta}$, we need to use a numerical maximization routine. If we are able to obtain the derivatives of $\ln(L)$ with respect to N and $\underline{\theta}$, then the Newton–Raphson method together with a Marquardt procedure can be used. This method also yields the Hessian matrix, which estimates the Fisher information matrix, and from which estimated variances of $\hat{N}$ and $\hat{\underline{\theta}}$ are obtained (see Section 2.4.1).

As before, the bootstrap yields more robust estimates of precision more simply, and we return to variance and interval estimation later.

Conditional MLE

Estimation is usually performed in two steps, using the conditional likelihood, rather than in one step from the full likelihood. An estimate $\hat{\underline{\theta}}$ of the parameter vector of the detection function is obtained by maximizing $L(\theta)$ (Equation (7.8)). This allows us to fit the detection function, and to estimate μ, the area under the function (call the estimate $\hat{\mu}$). We know that $E[p] = \mu/w$, so we can estimate $E[p]$ by

$$\hat{E}[p] = \hat{\mu}\pi_c \qquad (7.11)$$

We then use this in place of $E[p]$ in the encounter rate likelihood for N (Equation (7.2)), and maximize with respect to N, to get

$$\hat{N} = \frac{n}{\pi_c \hat{E}[p]} = \frac{n}{\frac{2wL}{A} \frac{\hat{\mu}}{w}} = \frac{nA}{2\hat{\mu}L} \qquad (7.12)$$

Density is estimated by

$$\hat{D} = \frac{n}{2\hat{\mu}L} \qquad (7.13)$$

This is the standard approach for line transect sampling, addressed in detail by Buckland *et al.* (2001).

Example: half-normal detection function

Technical
section
↓

Usually, we must again resort to numerical methods to maximize $L(\theta)$, but for the half-normal model with $w = \infty$, an analytic estimator of the single detection function parameter is easily obtained. For the half-normal model, $\underline{\theta}$ is simply the scalar σ^2 and $p(x) = \exp \frac{-x^2}{2\sigma^2}$. We write $E[p]$ explicitly in terms of the single parameter, as $E[p] = \frac{1}{w}\mu(\sigma^2)$, where

$$\mu(\sigma^2) = \int_0^\infty e^{\frac{-x^2}{2\sigma^2}} dx = \sqrt{\frac{\pi\sigma^2}{2}} \qquad (7.14)$$

The MLE of $\mu(\sigma^2)$ is

$$\hat{\mu} = \sqrt{\frac{\pi \sum_{i=1}^n x_i^2}{2n}} \qquad (7.15)$$

(see Exercise 7.2).

Using this estimate of μ, we estimate N by

$$\hat{N} = \frac{nA}{2\hat{\mu}L} = \frac{nA}{2L\sqrt{\frac{\pi \sum_{i=1}^n x_i^2}{2n}}} \qquad (7.16)$$

In the case of the St Andrews example data, $n = 88$, $A = 1,680\,\text{m}^2$, $L = 118\,\text{m}$ and $\sum_{i=1}^n x_i^2 = 442.04$. Thus we estimate that there are $\hat{N} = 210$ animal groups in the survey region. The true population size in this example was known to be 250 groups. In real problems, we do not know the true population size, and so it is important that we quantify the precision of our estimate.

↑
Technical
section

7.2.3 Horvitz–Thompson: estimation partly by design

The Horvitz–Thompson estimator was introduced in Chapter 4 (Equation (4.1)). For strip counts, it takes the form

$$\hat{N} \;=\; \sum_{j=1}^{J} \frac{N_{cj}}{p_{dj}} \tag{7.17}$$

where J strips are sampled, N_{cj} animals are counted in covered strip j, and p_{dj} is the inclusion probability (the probability that strip j is sampled), determined by the survey design. Clearly, if we sample all strips, we will have completed a full census, with $p_{dj} = 1$ for every j, and population abundance is simply the sum of our counts. If a simple random sample of strips is surveyed, and all strips are of equal size, then p_{dj} is simply the coverage probability $\pi_c = \frac{a}{A}$.

We can frame line transect surveys in similar terms but with the animals (or groups of animals), not the strips, as the sampling units. For line transect sampling, an animal appears in the sample if it is both within the covered area (i.e. one of the surveyed strips) and it is detected. We saw in Section 7.2.2 that this probability is $\pi_c E[p]$. (Recall that π_c is the probability that an animal is in the covered region and $E[p]$ is the average detectability of animals in the covered region.)

The Horvitz–Thompson estimator, given only that n animals were detected in the covered region, is therefore

$$\hat{N} \;=\; \sum_{j=1}^{J} \frac{1}{\pi_c E[p]} \tag{7.18}$$

Because $E[p]$ is unknown, we must estimate it, for example using the methods of Section 7.2.2. With line transect sampling, $E[p] = \frac{\mu}{w}$, so using the estimator $\hat{E}[p] = \frac{\hat{\mu}}{w}$ gives the "Horvitz–Thompson-like" estimator of abundance,

$$\hat{N} \;=\; \frac{n}{\frac{a}{A}\frac{\hat{\mu}}{w}} \tag{7.19}$$

With transects of half-width w and total length L, $a = 2wL$ and Equation (7.19) reduces to the conventional estimator for line transect sampling (Buckland et al., 2001):

$$\hat{N} \;=\; \frac{nA}{2\hat{\mu}L} \tag{7.20}$$

Although the Horvitz–Thompson estimator is unbiased, the "Horvitz–Thompson-like" estimator is not, unless $1/\hat{\mu}$ is an unbiased estimator of $1/\mu$. Note also that we have now introduced a modelling element to the design-based Horvitz–Thompson estimator, to allow us to estimate the inclusion probability $E[p]$.

We have used the distances of detected animals from the line to estimate μ, but in the above estimator this information was not used to give detection-specific inclusion probabilities. An animal at distance x_i has probability $p(x_i)$ of being detected, which we can estimate from our fitted model for the detection function. Thus an alternative Horvitz–Thompson estimator is

$$\hat{N} = \frac{A}{a} \sum_{i=1}^{n} \frac{1}{p(x_i)} \qquad (7.21)$$

This can also be expressed as $\hat{N} = \frac{A}{a} \sum_{j=1}^{J} \hat{N}_{cj}$, where $\hat{N}_{cj} = \sum_{i=1}^{n_j} \frac{1}{p(x_i)}$. The corresponding Horvitz–Thompson-like estimator is:

$$\hat{N} = \frac{A}{a} \sum_{i=1}^{n} \frac{1}{\hat{p}(x_i)} \qquad (7.22)$$

This estimator becomes unstable if many estimated detection probabilities are close to zero. The truncation distance w can be chosen to avoid this difficulty; a value such that $\hat{p}(w) \simeq 0.2$ should prove adequate.

With our example data, $\hat{p}(w) \approx 0.4$ and the point estimate of N using Equation (7.22) turns out to be the same as the conventional estimator of abundance, to the nearest integer. The variance of the estimate increases somewhat because $\hat{p}(x_i)$ is more variable than $\hat{E}[p]$. This formulation of the Horvitz–Thompson estimator opens up the way for modelling heterogeneity in the detection probabilities.

Suppose in addition to distance x_i, we record values $\underline{z}_i$ of several covariates for detection i. Then we can write the detection function as $p(x, \underline{z})$. If we specify and fit a model for this multivariate detection function (Marques and Buckland, 2003), then we can estimate abundance as

$$\hat{N} = \frac{A}{a} \sum_{i=1}^{n} \frac{1}{\hat{p}(x_i, \underline{z}_i)} \qquad (7.23)$$

or, with less variance, as

$$\hat{N} = \frac{A}{a} \sum_{i=1}^{n} \frac{1}{\int_0^w \hat{p}(x, \underline{z}_i) \pi(x)\, dx} = \frac{A}{a} \sum_{i=1}^{n} \frac{1}{\hat{E}_x[p(x, \underline{z}_i)]} \qquad (7.24)$$

7.2.4   Populations that occur in groups

Animals often occur in groups or clusters. This might be flocks of birds, schools of dolphins, or any other well-defined units. In this case, the distance of the group from the line is recorded, along with the group size. Full likelihood functions that include group size involve a state model for group size. We deal with these in the "Advanced Methods" section of the book, in Chapter 11.

Estimation by design

The form of the Horvitz–Thompson-like estimator in which summation is over individual detections (Equation (7.22)) readily extends to when the detections are of groups:

$$\hat{N} \;=\; \frac{A}{a}\sum_{i=1}^{n}\frac{z_i}{\hat{p}(x_i)} \tag{7.25}$$

Thus, for every group of size z_i that we detect, we estimate that there are $1/\hat{p}(x_i)$ in the covered region, together comprising $s_i/\hat{p}(x_i)$ animals. Again, using $\hat{E}[p(x)]$ instead of $\hat{p}(x_i)$ produces a less variable estimator. In fact, the form with covariates is more useful to us, as we can allow for size-bias in the detection probability if we have group size z as one of the covariates $\underline{z}$:

$$\hat{N} \;=\; \frac{A}{a}\sum_{i=1}^{n}\frac{z_i}{\hat{E}_x[p(x,\underline{z}_i)]} \tag{7.26}$$

where $\hat{E}_x[p(x,\underline{z}_i)] = \int_0^w \hat{p}(x,\underline{z}_i)\pi(x)dx$. If it is the only covariate in addition to distance x, we have

$$\hat{N} \;=\; \frac{A}{a}\sum_{i=1}^{n}\frac{z_i}{\hat{E}_x[p(x,s_i)]} \tag{7.27}$$

7.2.5   Interval estimation

If we use the full likelihood approach a parametric bootstrap could be applied by generating animal locations using a spatial state model (assumed uniform above) for the distribution of animals and using the fitted detection function to generate new detection distances. In common with analytic methods based on the information matrix, this has the disadvantages that detections must be assumed to be independent events, and that the variance estimates are conditional on the selected model for the detection function. We considered only independent uniform distribution of animals, but in reality this is often not a very appropriate model because animals or groups of animals tend to be more clustered, or less clustered if animals are territorial, than independent uniform distribution allows. These assumptions may be avoided by using the nonparametric bootstrap, in which the sampling unit is the individual line. We can ensure by design that data from different lines are independent, and the data of the resamples are not generated from an assumed model for the detection function. Note that this method of variance and CI estimation is design-based; it resamples the design units (the lines) and does not include the strong assumptions about animal distribution that are implicit in the likelihood function and parametric bootstrap. This gives more robust interval estimates.

If all lines are of equal length, then bootstrapping involves resampling lines (with replacement) until the resample contains the same number of lines as the original sample. If lines are of unequal length, this approach is again usually adopted, although resampling can continue until the total line length is equal, or nearly equal, to the total length of line in the original design. For reliable CI estimation, we would seek to have a design with at least 15–20 lines.

Thus the nonparametric bootstrap is implemented in much the same way as for plot sampling. If we condition on the number of lines, rather than on total line length, then:

(1) For replicate i $(i = 1, \ldots, B)$, choose J lines from those sampled, with replacement, where J is the number of sampled lines in the original design.

(2) Calculate the MLE or conditional MLE $\hat{N}_i$ for replicate i and store.

(3) Repeat steps (1) and (2) B times, where $B = 1,000$ is generally a reasonable choice.

(4) Calculate the mean and variance of the B $\hat{N}_i$s, as follows:

$$\hat{N}_b = \frac{\sum_{i=1}^{B} \hat{N}_i}{B} \tag{7.28}$$

$$\widehat{Var}_b\left[\hat{N}_b\right] = \frac{\sum_{i=1}^{B} \left(\hat{N}_i - \hat{N}_b\right)^2}{(B-1)} \tag{7.29}$$

(5) Estimate a 95% CI for N by $(\hat{N}_{(j)}; \hat{N}_{(k)})$, where $j = (B+1) \times 0.025$, $k = (B+1) \times 0.975$ and the notation $\hat{N}_{(j)}$ and $\hat{N}_{(k)}$ indicate the jth and kth largest $\hat{N}_i$s, respectively.

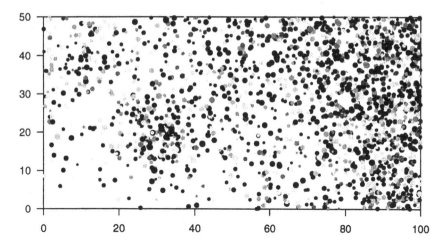

Figure 7.8. Simulated population of 2,000 groups. Circle size is proportional to group size, shading shows exposure: dark groups are easy to see, light groups are difficult to see.

The point estimate based on the original data, $\hat{N}$, is normally used in preference to the bootstrap mean, $\hat{N}_b$. $Var[\hat{N}]$ is estimated by $\widehat{Var}_b[\hat{N}_b]$.

7.2.6   An example with heterogeneity

Recall that unmodelled heterogeneity (differences in detectability due to animal-level variables) resulted in substantial bias when removal or mark-recapture methods were used to estimate abundance. One of the most attractive features of distance sampling methods is that estimates are very insensitive to unmodelled heterogeneity. This property is called "pooling robustness" in the distance sampling literature (Buckland et al., 2001).

To illustrate the application of line transect methods with more realistic data than we used above, and to illustrate the insensitivity to violation of the uniform distribution assumption and to heterogeneity in particular, we have used the R library WiSP to simulate a heterogeneous population of $N = 1,000$ groups from the spatial state model shown in Figure 3.3. Groups have a range of "exposures" that affect detection probability in such a way that the least exposed groups are about half as likely to be seen as the most exposed groups. The estimator ignores exposure altogether. The population is shown in Figure 7.8.

WiSP has further been used to simulate a line transect survey of the population, with a half-normal detection function that depends on group size and the 24 transects shown in Figure 7.9. The transect location was obtained using regularly spaced lines with a random start point. The detection function used in the simulated survey depends on animal exposure;

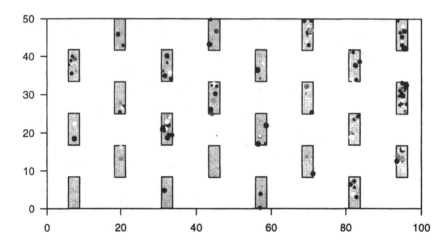

Figure 7.9. Location of searched strips (shaded) and transects (not shown, but running down the centre of the strips) and detected groups (dots) on a simulated survey of the simulated population of 2,000 groups.

that used for estimation ignores animal exposure entirely. In practice selecting an appropriate form of detection function is a key part of distance sampling survey analysis and AIC is one of the criteria commonly used. We do not have space to consider detection function model selection here; see Buckland *et al.* (2001).

A total of $n = 136$ groups were detected. Their perpendicular distance distribution is shown in Figure 7.10, together with the half-normal detection function fitted by maximizing the conditional likelihood of Equation (7.8). Only 12% of the survey region was covered and only about half the animals in the covered region were detected. The resulting point estimate of group abundance using Equation (7.16) is $\hat{N} = 1,750$ groups, which is close to the true abundance. (Essentially the same fit and estimates are obtained from the full likelihood Equation (7.10).)

With these simulated data (and using the conditional MLE approach), $\widehat{cv}[\hat{N}] = 15\%$ and the 95% CI for N, the number of groups, is estimated using a nonparametric bootstrap to be (1,177; 2,517). When we estimate variance assuming uniform distribution of groups in the survey region, $\widehat{cv}[\hat{N}] = 11\%$ and the 95% CI is estimated to be (1,370; 2,368). Using a state model which assumes a uniform distribution of groups gives negatively biased estimates of variance when the true distribution is more clustered than uniform (as it is here). Bootstrapping on transects gives more robust variance and confidence interval estimates.

The fact that $\hat{N}$ is so close to the true N for this single survey does suggest that the line transect estimator of N is not biased substantially by ignoring the heterogeneity due to animal exposure. This may be chance;

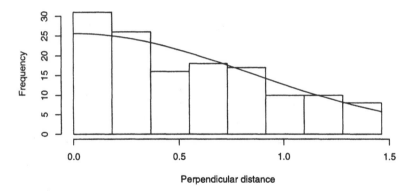

Figure 7.10. Detection function fit to the simulated survey of the simulated population of 2,000 groups. The smooth curve is the fitted half-normal detection function.

to investigate it more rigorously we simulated 500 surveys of this population. The distribution of the 500 $\hat{N}$ estimates is shown in Figure 7.11. The estimated bias in $\hat{N}$ is less than 1%, despite the presence of substantial heterogeneity; the line transect estimator is "pooling robust". Contrast this plot with the plots of simulated $\hat{N}$ distributions in the presence of heterogeneity in Chapters 5 and 6.

7.3 Point transect sampling

Point transect sampling is an extension of "point counts", which are a special case of plot sampling, in which the J plots are circles of radius w. The covered region has area $a = J\pi w^2$ (note that this π is the number $3.141593\ldots$, not a symbol for a state model). If n is assumed to be all animals within this region, $\hat{N} = nA/a$ (Chapter 4). For point transect sampling, some animals within the region are missed, and distances of detected animals from the point are used to model this added element.

Thus a point transect survey in its simplest form comprises J points randomly positioned throughout the study region, or more usually a systematic grid of J points randomly superimposed on the region. The observer visits each point, and records the detection distance x from the point to each detected animal. All animals on or very close to the point are assumed to be detected, and probability of detection is assumed to decrease with distance from the point, out to distance w.

A simulated point transect survey

To illustrate the method, we use WiSP to simulate a point transect survey of the same population used in the line transect simulation example above.

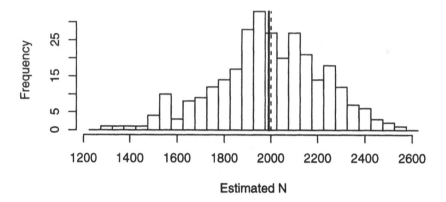

Figure 7.11. Distribution of $\hat{N}$ from 500 simulated surveys. The dotted vertical line is at the true N, the solid vertical line is at the mean of the 500 estimates.

Using the same detection function as was used in the line transect example, a truncation distance of $w = 3\,\mathrm{m}$, and 50 points, 96 groups were detected. The location of the points and detected groups is shown in Figure 7.12.

The distribution of radial distances of detected groups is shown in Figure 7.13, together with the fitted density function, $\hat{f}(x)$. Note that, unlike line transect survey data, the number of detections close to the point is low and initially increases with distance from the point before falling off again. This is not because detection probability near the point is low, but because the surface area near the point is small and there are therefore few animals in it to be detected. The area of concentric "doughnuts" about the point increases rapidly as you move away from the point and this accounts for the initial increase in numbers of detections. At some point (around 1 m in Figure 7.13), the drop in detection probability overcomes this increase and the frequency of detections begins to decline. The estimated detection function is shown in Figure 7.14. Note that the histogram bars in this figure do not correspond to detected frequencies; they have been scaled to take account of the differences in expected frequencies at different distances from the line. Goodness of fit of the detection function should not therefore be judged from Figure 7.14.

Group abundance is estimated to be 1,448 with a confidence interval of (871; 2,407), using the conditional maximum likelihood method and bootstrap variance estimation method described below. Again, our confidence interval comfortably includes the true abundance of $N = 2,000$ groups.

7.3.1 Maximum likelihood estimation

As with line transect sampling, to define a full likelihood framework, we need to specify a state model for the locations of all groups at the time of

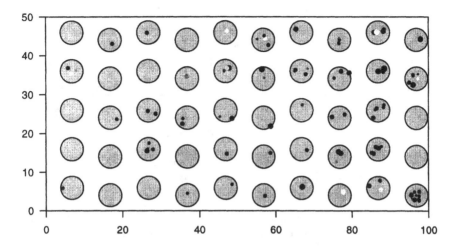

Figure 7.12. Location of point transects and detected groups on a simulated survey of the simulated population of 2,000 groups.

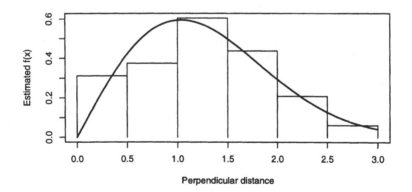

Perpendicular distance

Figure 7.13. Estimated probability density function, $\hat{f}(x)$ from the point transect survey of the simulated population of 2,000 groups.

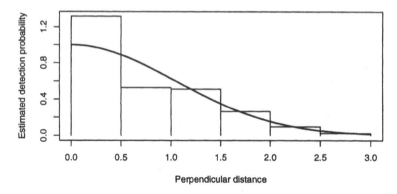

Figure 7.14. Estimated half-normal detection function from the point transect survey of the simulated population of 2,000 groups.

the survey. The essential difference between line and point transect sampling lies in the probability density of animals in the covered region with respect to distances from the line or point. In the case of line transects, it is reasonable to assume that the expected number of animals is the same at all distances from the line in the covered region. In the case of point transect surveys, the expected number of animals increases as distance from the point increases, because the surface area of concentric annuli about the point increases as you move away from the point.

State and observation models

The observation models for point transect sampling are the same as those for line transect sampling.

Suppose we again assume a uniform state model, in which a group is equally likely to be anywhere in the survey region of size A, so that the probability that it falls in a covered region of area a is a/A. Then the expected number of groups in an annulus comprising the area between distances $(x, x + dx)$ from the point is proportional to the surface area of the annulus. For small dx, this is to first order $2\pi x\, dx$. It follows that the probability density of radial distances of groups within a circle of radius w about the point is

$$\pi(x) \quad = \quad \frac{2\pi x}{\pi w^2} \quad = \quad \frac{2x}{w^2} \tag{7.30}$$

The fact that this density depends on the radial distance, x, makes the analysis of point transect surveys a little more complicated than the analysis of line transect surveys, in which $\pi(x) = \frac{1}{w}$.

MLE from the full likelihood

The derivation of a full likelihood for point transect sampling is identical to that for line transect sampling, except for the fact that $\pi(x) = \frac{2\pi x}{\pi w^2}$. Substitution in Equation (7.4) gives

$$E[p] = \int_0^w p(x)\frac{2\pi x}{\pi w^2}\, dx = \frac{\nu}{\pi w^2} \qquad (7.31)$$

where $\nu = 2\pi \int_0^w xp(x)\, dx$. Like $p(x)$, $E[p]$ is a function of the parameter vector $\underline{\theta}$.

With this change from the line transect case, the full likelihood is

$$L(N;\underline{\theta}) = \binom{N}{n}(1 - \pi_c E[p])^{N-n}(\pi_c)^n \prod_{i=1}^{n} p(x_i)\frac{2x_i}{w^2} \quad (7.32)$$

The same numerical maximization methods may be used as for line transect sampling.

Conditional MLE

We can adopt a similar strategy to line transect sampling and condition on sample size n. Substitution into Equation (7.8) gives the conditional likelihood

$$L(\underline{\theta}) = \left(\frac{2\pi}{\nu(\underline{\theta})}\right)^n \prod_{i=1}^{n} x_i p(x_i) \qquad (7.33)$$

As for line transect sampling, we can now maximize this likelihood with respect to the $\underline{\theta}$, and hence obtain conditional maximum likelihood estimates of the parameters of the detection function. This gives us the MLE of ν, which we denote $\nu(\hat{\theta})$ or more briefly, just $\hat{\nu}$.

We now need to estimate N. If w is finite, the covered region has area $J\pi w^2$, where J is the number of points. Within this area, n animals were detected. We can estimate N using Equation (7.12) and the appropriate $\hat{E}[p]$ for point transect surveys:

$$\hat{N}_c = \frac{n}{\left(\frac{\hat{\nu}}{\pi w^2}\right)} \qquad (7.34)$$

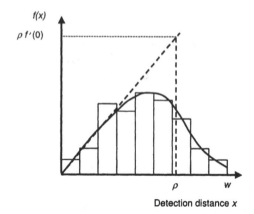

Detection distance x

Figure 7.15. The effective radius ρ is the distance for which as many animals are detected beyond ρ as are missed within ρ. The diagonal dotted line shows the slope of the density $f(x)$ at distance zero, which is denoted $f'(0)$. Because the area under $f(x)$ is unity, and equals the area of the triangle, we have $\rho^2 f'(0)/2 = 1$, or $\rho = \sqrt{2/f'(0)}$.

As in line transect sampling, we may allow $w \to \infty$, so that the result holds whether or not w is finite. Density is estimated by

$$\hat{D} = \frac{n}{J\hat{\nu}} \qquad (7.35)$$

This is the standard approach for point transect sampling, given by Buckland et al. (2001). If we define $\nu = \pi\rho^2$, then ρ is the effective radius of the circle, in the sense that the expected number of animals detected at distances $> \rho$ from the point equals the expected number missed within ρ of the point (Figure 7.15). The effective area covered is $J\nu = J\pi\rho^2$. In line transect sampling, the probability density function of distances of detected objects from the line is proportional to the detection function $p(x)$. In point transect sampling, it is proportional to $x\,p(x)$ because area in an annulus at distance x from the point increases linearly with x.

↑
Technica
section

7.3.2 Horvitz–Thompson: estimation partly by design

We need to make only minor changes to the development for line transect sampling. Using $\hat{E}[p] = \frac{\hat{\nu}}{\pi w^2}$ and $a = J\pi w^2$ (where J is the number of points) in Equation (7.18), we get the Horvitz–Thompson-like estimator of abundance for point transect surveys

$$\hat{N} = \frac{nA}{J\hat{\nu}} \tag{7.36}$$

This is the conventional estimator for point transect sampling (Buckland et al., 2001). The point transect Horvitz–Thompson-like estimator that uses the estimated individual detection probabilties in place of $\hat{E}[p]$ can be written

$$\hat{N} = \frac{A}{a} \sum_{i=1}^{n} \frac{1}{\hat{p}(x_i)} \tag{7.37}$$

We can also model the effects of heterogeneity in the detection probabilities as for line transect sampling. Given covariates $\underline{z}$, the detection function is $p(x, \underline{z}) \equiv p(x, \underline{z}; \underline{\theta})$. We specify and fit a model for this multivariate detection function, and estimate abundance as

$$\hat{N} = \frac{A}{a} \sum_{i=1}^{n} \frac{1}{\hat{E}_x[p(x, \underline{z}_i)]} \tag{7.38}$$

We discuss this sort of estimator more in Chapter 11.

7.3.3 Populations that occur in groups

The full likelihood for populations that occur in groups requires specification of a state model for group size. We deal with this in Chapter 11.

A Horvitz–Thompson-like estimator without size bias is

$$\hat{N} = \frac{A}{a} \sum_{i=1}^{n} \frac{s_i}{\hat{E}[p]} \tag{7.39}$$

where s_i is the size of the ith group. If we have covariates $\underline{z}$, one of which might be group size, to allow modelling of size bias, we have

$$\hat{N} = \frac{A}{a} \sum_{i=1}^{n} \frac{s_i}{\hat{E}_x[p(x, \underline{z}_i)]} \tag{7.40}$$

7.3.4 Interval estimation

This proceeds much as for line transect sampling. If the points are randomly located through the study region, then the nonparametric bootstrap is applied by resampling points (along with the distances associated with them) with replacement until the resample has the same number of points as the original sample. If the design comprises a regular grid of points, spanning the entire study region, we assume that these represent a random sample of points, and resample as before. Sometimes, point transect surveys are designed by placing parallel, equally spaced lines through the study region, and placing points at regular intervals along the lines. Unless the gap between successive lines is equal to the gap between successive points on a line (in which case a regular grid of points is obtained), the *lines* should be treated as the sampling units when generating resamples. That is, lines are resampled; if a line is selected, all points on that line, and their associated data, are included in the resample. In this case, within a resample, resampling might stop once the same number of lines has been selected as for the original design, or alternatively, it might stop as soon as the number of points is greater than or equal to the number in the original sample. In the latter case, the last line selected would be dropped if by doing so, the remaining number of points is closer to that for the original sample.

A nonparametric bootstrap gives an estimated 95% CI of (871; 2,407). If variance estimation is done on the basis of a state model with uniform group location (which, from Figure 7.8, is clearly not appropriate), the CI is estimated to be (1,069;1,961). As expected, the assumption of uniform spatial state model results in a CI that is too narrow. The effect is less marked than in the line transect example above, partly because in point transect sampling, estimation of the detection function contributes a larger component of the variance than it does with line transects; the inappropriate state model affects primarily the estimate of encounter rate variance (the variance in the number of groups per unit survey area). In the simulated line transect survey, the encounter rate contributed 59% of the variance of $\hat{N}$, while in the simulated point transect survey, it contributed only 38%. Negative bias in encounter rate variance resulting from an inappropriate state model therefore has less effect on the overall variance in the case of point transects.

The estimation of the detection function in point transect surveys typically involves substantially more uncertainty than in line transect surveys. It is the data close to distance $x = 0$ that are most important for estimating the effective strip width/circle radius, but in the case of point transect surveys, the region close to $x = 0$ is the region in which least data are gathered, because so little area is close to $x = 0$.

7.4   Other distance sampling methods

We consider briefly three further distance sampling methods. The first, cue counting, has been used exclusively for marine mammal surveys carried out by ship or aircraft. The number of cues (usually whale blows) per unit area per unit time is estimated, and in a separate experiment, the cue rate of individual animals is estimated. By dividing the first estimate by the second, an estimate of animal density is found. The second method is trapping webs, in which sample points are chosen, and at each point, a web of traps is laid out, with higher trap density close to the web centre. Trapped animals provide the detections, and the detection distance is the distance of the trap that captured the animal from the web centre. The third method, indirect distance sampling, is designed for animals that cannot themselves be satisfactorily surveyed, but that produce signs that can be. The two most common examples are nests and dung. The surveys yield estimates of nest or dung density, and nest or dung decay rates and production rates are separately estimated, to allow conversion from nest or dung density to animal density.

Further details of all three methods are given by Buckland *et al.* (2001).

7.4.1   Cue counting

Suppose a shipboard survey of whales is conducted, in which observers scan a sector of angle γ ahead of the ship, recording any detected whale blows within the sector, and their distances from the ship (Figure 7.16). This is similar to point transect sampling, in which a proportion $\gamma/2\pi$ of the circle is counted. However, instead of sampling J points, the single sector is monitored for a time T. If animal density were uniform, the ship need not move, but in practice, the method is far more robust if the survey has good spatial coverage of the survey region, achieved by travelling along transects much as for line transect sampling.

Assuming that cues occur uniformly in time, and independent of the location of the ship, then the full likelihood for point transect sampling (Equation (7.32)) applies to cues, except that, because cues are (assumed to be) instantaneous, the coverage probability is the sector area $\gamma\pi w^2/2\pi$ divided by the area of the survey region A: $\frac{\gamma w^2}{2A}$.

Conditional maximum likelihood estimation of the parameters of the detection function follows exactly as for point transect sampling (Section 7.3.1). Because the covered area at any instant is the sector area, we must multiply by time T to obtain the total covered area, $\gamma T w^2/2$. (It does not matter that these areas overlap in space, as cues are instantaneous. The units of covered "area" are in fact area-time.) We therefore estimate the number of cues per unit area per unit time by

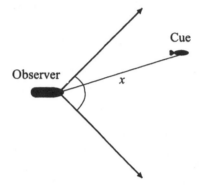

Figure 7.16. In cue counting, an observer travels along a transect line, recording all detected cues within a sector of angle γ ahead of the observer. Cues are assumed to be discrete moments in time at which an animal is detectable, such as occurs when a whale blows, and the same animal may be recorded several times, if it gives several cues whilst in the sighting sector.

$$\hat{D}_c = \frac{n}{\gamma T w^2/2 \times \frac{\hat{\nu}}{\pi w^2}} = \frac{2n\pi}{T\gamma\hat{\nu}} \tag{7.41}$$

To convert this estimate into an estimate of animal abundance, we must divide by an estimate of the average cue rate of the population η (cues per animal per unit time) and multiply by the size of the survey region:

$$\hat{N} = \frac{2n\pi A}{T\gamma\hat{\nu}\hat{\eta}} \tag{7.42}$$

Estimation of η can of course be accomplished by extending the full likelihood, to include a component corresponding to a cue rate estimation experiment. In practice, it is simpler to estimate η by the sample mean of the individual cue rates observed for the animals monitored in the experiment.

Similar modifications can be made to the Horvitz–Thompson estimators for point transects in Equations (7.37) and (7.38). Interval estimation can be achieved using the nonparametric bootstrap with transect lines as the sampling units, as for line transect sampling. Thus the lack of independence between cues from the same animal or group of animals do not invalidate this confidence interval.

7.4.2 Trapping webs

Trapping webs comprise a sample of circular plots. On each plot, a web of traps is located, with close spacing near the centre and progressively

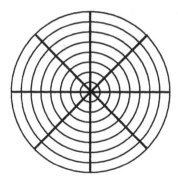

Figure 7.17. A trapping web comprises a number of traps (65 in this example, one at each intersection of a "spoke" and a circle), spaced so that all animals close to the web centre will be caught. Trap spacing increases with distance from the centre, so that probability of capture falls off in much the same way as probability of detection falls off with distance from the point in point transect sampling.

wider spacing as distance from the centre increases (Figure 7.17). The philosophy is that trap spacing should be sufficiently small to ensure that all animals close to the web centre are trapped. The distance data are the distances from the web centre to the traps in which the "detected" animals are caught. This is therefore like point transect sampling, and exactly the same likelihoods and estimators apply.

A weakness of the trapping web method is that it depends on movement of the animals in order to trap them, yet the concept behind point transect sampling is that the locations of animals are recorded as if at an instant in time. Because trap density is higher near the centre of the web, animals that move over a large area are more likely to be trapped close to the centre, causing upward bias in the density estimates.

7.4.3   Indirect distance sampling surveys

In indirect distance sampling surveys, the line or point transect survey is conducted not on the animal of interest but on the signs it produces. This is usually dung-piles (e.g. elephants), dung pellet groups (e.g. deer), or nests (e.g. apes). Here, we suppose that dung-piles are surveyed. Standard line or point transect methods apply to the dung-pile survey, so the likelihoods and estimators of Section 7.2 or 7.3, depending on which technique was used, apply. Having estimated dung-pile density, it must be divided by an estimate of the mean time to decay (in days) of dung-piles in the period preceding the survey, which yields an estimate of the number of dung-piles deposited per day. To convert this to an estimate of animal density, we need to divide again by an estimate of the number of dung-piles produced per day per animal.

Thus experiments are needed to estimate mean time to decay and defecation rate. If these parameters are modelled in a likelihood framework, the full likelihood may be obtained by multiplying that for line transect sampling by the likelihoods corresponding to the decay rate and defecation rate experiments. Because the three components of the survey are independent, for most purposes, it is simpler to model them independently, and then use for example the delta method (Seber, 1982) to estimate the variance of the composite estimator.

7.5   Effect of violating main assumptions

Assumption 1 All animals at distance zero are detected: $p(0) = 1$.[1]

Effect of violation: Estimates of abundance are negatively biased in proportion to $p(0)$. For example, if $p(0) = 0.25$, estimates will on average be only 25% of the true abundance.

Assumption 2 Groups are randomly (uniformly) and independently distributed in the survey region.

Effect of violation: Confidence intervals based on the assumption tend to be biased. Provided robust interval estimation methods are used (e.g. transect-based nonparametric bootstrap), violation of this assumption is of no great consequence.

Assumption 3 Animals do not move before detection.

Effect of violation: Random movement induces positive bias (the encounter rate is inflated). Provided object movement is slow relative to movement of the observer, the bias is small. Responsive movement can cause large bias (positive if there is attraction to the observer, negative if there is avoidance).

Assumption 4 Distances are measured accurately.

Effect of violation: Line transect estimators are fairly robust to random errors in measurement; point transect estimators are more sensitive. They are both sensitive to systematic bias in distance measurement, and to rounding to zero distance.

Providing assumption 1 holds, distance sampling estimators are pooling robust. That is, no bias is introduced by pooling data from animals with

---

[1]This assumption is referred to as the "$g(0) = 1$" assumption in the distance sampling literature, because in this context detection probability is conventionally written as $g(x)$, not $p(x)$.

different detection probabilities. This is a powerful feature of distance sampling methods because animal populations are almost always heterogeneous and modelling this can be difficult.

7.6   Summary

The methods considered in the preceding chapters involved either a state model (plot surveys) or an observation model (removal methods and mark-recapture methods), but not both. Distance sampling involves both a state model (for distances) and an observation model (probability of detection, given distance).

The assumption that detection of animals in the observer's path is certain, allows estimation of abundance without repeat surveys. The rate at which detection frequency per unit area searched falls off with distance contains information on the relative detectability of animals with distance, but without knowing detection probability at (at least) one distance, the fall-off rate contains no information about absolute detectability.

Distance sampling state models are unusual because they are known (no state model parameters are estimated). They imply that animals are uniformly distributed in the survey region. This is usually not very plausible at the scale of the survey region, but over the short distances within w of the point or line, it is often quite reasonable. Design-based interval estimation methods like a transect-based bootstrap, are likely to give more reliable interval estimates than those from a full likelihood involving an assumption of independently uniformly distributed animals throughout the survey region.

With removal methods and mark-recapture methods, the presence of unmodelled heterogeneity is a cause for concern because it can lead to large bias. Distance sampling methods are robust to unmodelled heterogeneity. This property, called "pooling robustness" in the distance sampling literature, is a very desirable property because animal populations are almost always heterogeneous (i.e. not all individuals are equally detectable), and modelling heterogeneity adequately can be difficult.

In Part III, we consider distance sampling and other methods with more general state models than we have dealt with thus far. This includes state models for location (spatial modelling) as well as for sources of heterogeneity like size and sex. Horvitz–Thompson-like estimators of the sort introduced in this chapter turn out to be useful in quite a variety of contexts when heterogeneity is modelled. Before we deal with these more advanced estimation methods, however, we consider very briefly nearest-neighbour and related methods.

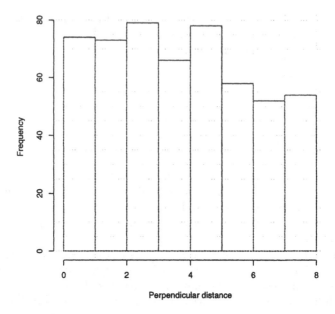

Figure 7.18. The distribution of perpendicular distances (in m) of duck nests observed on a line transect survey.

7.7   Exercises

Exercise 7.1   Figure 7.18 shows the numbers of duck nests detected on a line transect survey. A total of 1,000 km of transect were surveyed and 534 nests were detected. Perpendicular distances are measured in metres.

(a) Fit a smooth detection function to these data by eye.

(b) Use the curve to estimate the average probability of detecting a nest that is within 8 m of the transect line ($\hat{E}[p]$). Hence estimate the number of nests within 8 m of the transect line.

(c) If the survey area was 64 km², what is your estimate of N?

Exercise 7.2   Show that when the detection function has half-normal form on a line transect survey, and $w \to \infty$, the conditional MLE of $\mu \equiv \mu(\sigma^2)$, obtained from likelihood Equation (7.8), is

$$\hat{\mu} = \sqrt{\frac{\pi \sum_{i=1}^{n} x_i^2}{2n}} \qquad (7.43)$$

Exercise 7.3   Show that when the detection function has half-normal form on a point transect survey, and $w \to \infty$, the conditional MLE of ν, obtained from likelihood Equation (7.33), is

$$\hat{\nu} = \frac{\pi \sum_{i=1}^{n} x_i^2}{n} \qquad (7.44)$$

and hence that N is estimated using Equation (7.37) by

$$\hat{N} = \frac{n^2 A}{J\pi \sum_{i=1}^{n} x_i^2} \qquad (7.45)$$

Exercise 7.4   Suppose perpendicular distances $x_1, \cdots, x_n$ are recorded during a line transect survey. The surveyor wants to fit the following negative exponential detection function to the data.

$$g(x) \quad = \quad \exp(-\lambda x), \; 0 \le x < \infty \qquad (7.46)$$

(a) Derive the maximum likelihood estimator $\hat{f}(0)$ of $f(0)$, the density function of perpendicular distances evaluated at distance zero.

(b) Using the results of Appendix C, derive an estimator for the variance of $\hat{f}(0)$.

Exercise 7.5   In point transect sampling, suppose detection distances are squared prior to analysis.

(a) Derive the appropriate estimator of D, analogous to Equation (7.13), for use on these squared distances.

(b) Why does the result of part (a) allow line transect software to be used to analyse point transect data? Can you see a disadvantage of this approach?

Exercise 7.6   In a field test of line transect sampling, a known number of objects, N_c, were distributed uniformly through a surveyed strip of half-width w and length L, so that the covered area was $a = 2wL$. Two teams independently surveyed the strip. Both detected exactly half of the objects present, and recorded the distances from the line of the objects detected. These distances were analysed using conventional line transect methods, and $E(p)$, the proportion of objects detected, was estimated for each team.

The first team obtained $\hat{E}(p) = 0.54$, close to the true value. However, the second team obtained $\hat{E}(p) = 0.98$. Explain how this is possible, and sketch a detection function for each team that is consistent with these results.

Exercise 7.7   Consider a line transect survey in which distances are recorded in just two categories, within or beyond a specified distance x_0. Suppose n animals are detected from lines of total length L, of which n_0 are within distance x_0 of the line. Assume that probability of detecting a bird at distance x is given by $p(x) = \exp(-\lambda x), 0 \leq x < \infty$.

(a) Given that an animal is detected, what is the probability that it is within x_0 of the line? (Express your answer as a function of λ.)

(b) Hence write down the probability function $p(n_0)$ of n_0.

(c) Show that the maximum likelihood estimator of λ is given by $\hat{\lambda} = -\frac{1}{x_0} \log_e \left(1 - \frac{n_0}{n}\right)$.

(d) Hence obtain an estimator of animal density, D.

(e) Describe how you would estimate the variance of your density estimate (i) analytically and (ii) by a computer-intensive method.

(f) Is the above model a reasonable choice for $g(x)$? Explain your answer, and suggest an alternative model that might be better.

(g) Discuss the advantages and disadvantages of collecting data in just two distance intervals. You should consider both practical field issues and any problems that such a survey creates for the data analyst.

(h) Repeat parts (a)–(d) if the survey was a point transect survey, the detection function was half-normal with $p(x) = \exp\{-\lambda x^2\}, 0 \leq x < \infty$, and of n animals detected, n_0 were within distance x_0 of a point.

8

Nearest neighbour and point-to-nearest-object

Key idea: when animal density is high, distances to the nearest animals are shorter than when density is low.

Number of Surveys:	Only one survey required.
State Model:	Animals are assumed to be distributed uniformly and independently in the survey region.
Observation Model:	Nearest object is detected with certainty.

8.1 Introduction

In nearest neighbour sampling, objects within the survey region are randomly sampled, and the distances to their nearest neighbour are measured. In point-to-nearest-object sampling, points within the survey region are randomly sampled, and the distance from each point to the nearest object is measured. If we assume that objects are independently and uniformly distributed throughout the survey region, then the distribution of nearest-object distances is the same under both approaches.

Likelihood function and key notation:

$$L(N) = \left[\frac{2\pi N}{A}\right]^{J} \prod_{j=1}^{J} x_j \exp\left\{-\frac{\pi N x_j^2}{A}\right\}$$

N: number of animals in the population.
J: number of points/objects sampled.
A: surface area of whole survey region.
x_j: distance from jth point/object to nearest object.

The methods readily extend to the case that we measure the distance to the K nearest objects for $K \geq 1$ (Diggle, 1983). In areas of high density, $K > 1$ will give more efficient estimation of density.

The methods of this chapter are seldom used for practical wildlife management for several reasons:

- If objects are sparsely distributed in at least some of the area, it may prove difficult and costly to locate the nearest object.

- The methods are inefficient because many other objects may be located while trying to identify the nearest object; recording the distance to the k nearest objects may not help, as many more objects may be located while determining which are the k nearest objects.

- If objects are mobile, measurement of the nearest-object distance may be bias-prone; for this reason, the method is better suited to objects that stay in one place, such as plants.

- For the nearest neighbour method, it is difficult to select a random sample of objects from the entire population.

- The state model, which specifies that objects are uniformly distributed throughout the survey region, is unrealistic for most applications, and density estimation is very sensitive to this assumption.

The last problem has long been recognized, and various attempts have been made to adjust for non-uniformity. Indeed, the methods have more commonly been used to measure non-uniformity than to estimate abundance. If the distribution tends to be more clustered or aggregated than a random pattern, abundance is underestimated if distances are measured from a random point, and overestimated if measured from a random object. Some *ad hoc* estimators are therefore based on an average of the two estimates (Diggle, 1983).

8.2   Maximum likelihood estimation

Superficially, point-to-nearest-object surveys appear similar to plot sampling. Conceptually, plots are sampled from the survey region, where each plot is a circle about the sampled point with radius equal to the distance to the nearest object. The complication is that the size of the plot is now a random variable, not a known constant. This rules out the use of design-based methods.

We need therefore to find a likelihood function, given our assumed state model. We do this by deriving a pdf for distance x to the nearest object (or equivalently, for plot area, if plot is defined as above). This derivation applies for both nearest neighbour and point-to-nearest-object methods. It is based on the assumption of a uniform spatial state model, i.e. $\pi(x) = \frac{1}{A}$, as in plot sampling (Chapter 4).

Technical section ↓

Let X be the random variable "distance to nearest object", and x be a particular value this random variable takes.[1] The probability that the random variable X is greater than the value x, $Pr(X > x)$, is the probability that no objects occur within an area of size πx^2.[2] Given that objects are independently and uniformly distributed throughout the survey region, i.e. that the state model for an animal's location given by Cartesian coordinates (u, v), is $\pi(u, v) = \frac{1}{A}$, the number in an area of unit size has a Poisson distribution with rate N/A. Thus the number in an area of size πx^2 has a Poisson distribution with rate $\pi x^2 N/A$. Hence

$$Pr(X > x) \quad = \quad \frac{\left(\frac{\pi x^2 N}{A}\right)^0 \exp\left\{-\frac{\pi x^2 N}{A}\right\}}{0!}$$

$$= \quad \exp\left\{-\frac{\pi x^2 N}{A}\right\} \qquad (8.1)$$

Thus the cumulative distribution function of x is

$$F(x) \quad = \quad Pr(X \leq x) = 1 - \exp\left\{-\frac{\pi x^2 N}{A}\right\} \qquad (8.2)$$

---

[1] In this section we need to revert to standard statistical notation for random variables (capitals) and values of random variables (lowercase letters); in most of the rest of the book we do not make this distinction explicitly.

[2] $\pi(\ )$ indicates a probability density function, π is the ratio of a circle's circumference to its radius; do not confuse the two.

Differentiating with respect to x, we obtain the pdf, which is our state model for x:

$$\pi(x) \;\; = \;\; \frac{2\pi x N}{A} \exp\left\{-\frac{\pi x^2 N}{A}\right\} \qquad (8.3)$$

To obtain the likelihood function, we evaluate this for each of the J nearest-object distances, and multiply the evaluations together:

$$
\begin{aligned}
L(N) \;\; &= \;\; \prod_{j=1}^{J} \pi(x_j) \\
&= \;\; \left[\frac{2\pi N}{A}\right]^{J} \prod_{j=1}^{J} x_j \exp\left\{-\frac{\pi N x_j^2}{A}\right\} \qquad (8.4)
\end{aligned}
$$

It is now a simple matter to take logs, differentiate with respect to the unknown parameter N, and set equal to zero. Solving for N yields the estimator

$$\hat{N} = \frac{JA}{\pi \sum_{j=1}^{J} x_j^2} \qquad (8.5)$$

If we measure instead to the K nearest objects, the derivation yields

$$\hat{N} = \frac{JKA}{\pi \sum_{j=1}^{J}\sum_{k=1}^{K} x_{jk}^2} \qquad (8.6)$$

where x_{jk} is the distance to the kth nearest object for the jth sampled point or object.

↑
Technical
section

8.3   Interval estimation

The parametric bootstrap may be implemented by substituting $\hat{N}$ for N in the pdf of x (Equation (8.3)) and generating J deviates from it. (This task is made a little easier if we transform to plot area, as its pdf is the standard exponential distribution, from which it is easy to generate deviates.)

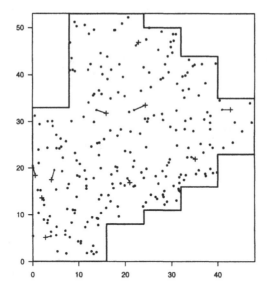

Figure 8.1. Example dataset, showing ten randomly located points (the crosses) within the survey region and distances to the animals nearest to each point (the lines).

This resample is analysed using Equation (8.5), and the process repeated a large number of times. Variance estimates and confidence intervals are then obtained as usual.

The nonparametric bootstrap proceeds similarly, but the resamples are obtained by selecting J points (and their associated distances) with replacement from the J in the original sample. If the design of a point-to-nearest-object survey is such that points are located along randomly or systematically placed lines, then the lines are resampled instead of the points.

8.4   Example

We illustrate only the point-to-nearest-object method; the nearest-neighbour method performs similarly but the bias caused by clustering is in the opposite direction (as noted above).

Figure 8.1 shows a sample of ten randomly chosen points in the example dataset survey region, together with the animals closest to them. With these data, Equation (8.5) gives an estimate of 201 objects, with a nonparametric bootstrap 95% confidence interval of (133; 365). The true population size is 250, so the estimator seems to have done quite well. This is because the distribution of animals is close to uniform.

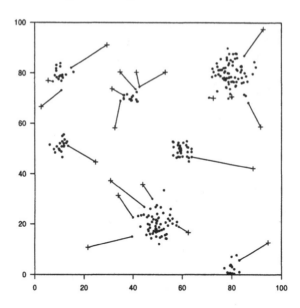

Figure 8.2. Simulated clustered population of 250 animals, showing twenty randomly located points (the crosses) and distances to the animals nearest to each point (the lines).

How does it perform when the uniform spatial state model assumption is violated? Figure 8.2 shows a very clustered population of 250 animals, which was generated using the R library WiSP. Twenty points were chosen at random in the survey region. This gave a point estimate of 22 animals, with a nonparametric bootstrap 95% confidence interval of (15; 36). The figure makes the source of the bias apparent: the distances from the points to the nearest animals are much larger on average than would be the case were the population uniformly distributed through the survey region. Hence the denominator of Equation (8.5) is too large and $\hat{N}$ is too small. Note that the opposite would be true for the nearest neighbour estimator; in this case all distances would be between points within the clusters.

To illustrate how biased the point-to-nearest-object estimator is with this population, we simulated 500 point-to-nearest-neighbour surveys of the population, each with 20 points, using WiSP. The simulated distribution of the estimator is shown in Figure 8.3. The mean of the simulated estimates is 19, which implies the estimator underestimates the number of animals in the population by about a factor of 13. This is huge negative bias!

Although the population we simulated here was very clustered, it does serve to illustrate the danger of using these methods when the population is not known to be approximately uniformly distributed in the survey region. Bias decreases as the level of departure from a uniform spatial state models decreases. For example, simulated 20-point surveys of a population of 2,000

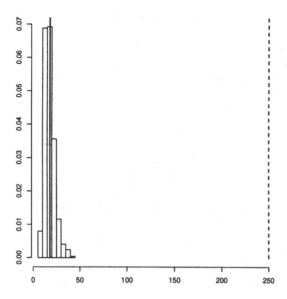

Figure 8.3. Distribution of the point-to-nearest-object MLE from 500 simulated surveys of the clustered population. The solid vertical line is the mean of the estimates; the dashed vertical line is the true population size.

animals generated from the state model shown in Figure 3.3 indicate that the estimator is negatively biased by 40% in this case. This is still unacceptably large bias. The R library WiSP can be used to experiment with alternative scenarios and to get a feel for the sensitivity of the methods to violation of the uniform spatial state model assumption.

We have only presented confidence interval estimates using the nonparametric bootstrap method here. This is because for aggregated populations, we can expect the nonparametric bootstrap to give more reliable confidence intervals than asymptotic normal, profile likelihood, or parametric bootstrap intervals because the resamples are generated without assuming that objects are uniformly distributed (although the subsequent estimation of abundance from the resamples does assume this).

8.5 Summary

Although similar to plot sampling, for the methods of this chapter, plot size is a random variable. This means that the natural way to analyse nearest distance data is to specify a state model that defines the distribution of this random variable. A state model specified in terms of nearest distances is easily obtained from a uniform spatial state model specified in terms of Cartesian coordinates (as with plot sampling). This provides us with a likelihood function that depends only on the state model, and that can be readily maximized to estimate abundance.

When the spatial distribution of animals is not uniform, (a) it is not usually known, (b) obtaining a state model for nearest distances is more difficult, (c) the distribution will in general be different at different points in the survey region, and (d) the estimates of N depend heavily on the state model assumed. This makes nearest neighbour and point-to-nearest-object methods difficult to apply with nonuniform spatial state models. These methods are seldom used in practice. By contrast, with sensible designs, the point estimates from the other two methods covered thus far that involve state models (plot sampling and distance sampling) are quite robust to departures from the assumed uniform spatial state model. The associated interval estimators are not, but design-based methods, for which the assumption of a uniform distribution of objects is not required, can be used to give more robust interval estimates. Nearest neighbour and point-to-nearest-object methods are not amenable to design-based estimation because the size of the sample unit (the circular plot) can't be controlled by the surveyor, but depends on the animal distribution.

8.6   Exercises

Exercise 8.1   Suppose instead of X, the distance to the nearest object, we work with Y, the area to the nearest object, where $Y = \pi X^2$. If N objects are uniformly distributed through the survey region of size A, and J points are sampled at random,

(a) derive the likelihood $L(N)$

(b) show that the maximum likelihood estimator of N is given by

$$\hat{N} \;=\; \frac{JA}{\sum_{j=1}^{J} y_j} \tag{8.7}$$

Exercise 8.2   Suppose objects are distributed through the survey region according to an inhomogeneous Poisson process. That is, the process that generated object locations is Poisson, but with rate varying by location.

(a) Given large samples, would the mean distance from a random point to the nearest object be greater than, less than or about the same as the mean distance from a random object to its nearest neighbour?

(b) Suggest a test of the null hypothesis that the Poisson process is actually homogeneous, given two samples, one comprising random point to nearest object distances, and the other random object to nearest neighbour distances.

Exercise 8.3   Diggle (1983) gives the following *ad hoc* estimators of object abundance, given distances $x_1, \ldots, x_J$ from J random points to the nearest objects, and the corresponding shortest distances $y_1, \ldots, y_J$ from these objects to their nearest neighbours (subject to the constraint that the line joining the point to the nearest object should make an angle of at least 90° with the line joining that object to its nearest neighbour).

$$\hat{N} = \frac{2JA}{\pi \left(\sum_{j=1}^{J} x_j^2 + 0.5 \sum_{j=1}^{J} y_j^2 \right)} \tag{8.8}$$

$$N^* = \frac{JA}{\pi \sqrt{0.5 \sum_{j=1}^{J} x_j^2 \sum_{j=1}^{J} y_j^2}} \tag{8.9}$$

$$\tilde{N} = \frac{J^2 A}{2\sqrt{2} \sum_{j=1}^{J} x_j \sum_{j=1}^{J} y_j} \tag{8.10}$$

The first estimator is in fact the MLE, if the objects are uniformly distributed. The second is more robust to failures of this assumption, and the third is less sensitive to the occasional large x_j in a strongly aggregated pattern.

The survey shown in Figure 8.1 with $A = 1,680$ and $J = 10$, gives $\sum x_j = 14.8$, $\sum x_j^2 = 26.7$, $\sum y_j = 17.6$ and $\sum y_j^2 = 47.6$. The survey shown in Figure 8.2 with $A = 10,000$ and $J = 20$, gives $\sum x_j = 208$, $\sum x_j^2 = 2915$, $\sum y_j = 62$ and $\sum y_j^2 = 502$.

(a) Evaluate the estimators for each survey, and assess their performance, given that the true abundance is 250 animals in both cases.

(b) How could you estimate the confidence interval for N in each case?

Part III

Advanced Methods

9
Further building blocks

9.1 Introduction

We concentrate on model-based methods of estimating abundance in this book. There are advantages and disadvantages to doing this. If you were interested in the process determining the population state (the spatial distribution of animals, for example), this would be a reason to use model-based methods. Model-based methods can also lead to substantial improvement in precision, because some of the variation in density is "explained" by the model instead of being assigned to variance. The main reason you might not want to use model-based inference is that you never know the true process determining the population state, and if the state model you assume is not a good approximation to it, inferences may be biased. With the availability of flexible state models and adequate diagnostics, this is much less of a problem than it used to be, although there are still sometimes difficulties. In particular, the independence assumptions implicit in many state models can be unrealistic, and modelling dependence can be difficult. Unless care is taken, this can result in model-based estimates of variance that are too small and corresponding confidence intervals that are too narrow. A compromise approach is to use model-based methods for point estimation, and nonparametric, and essentially design-based methods for interval estimation.

State models were largely absent from the methods and models considered in Part II. We assumed uniform distribution of animals in Chapters 4 and 7 but did not consider state models for anything other than animal distribution. State models play a much more central role in the advanced methods we consider in this part of the book. Spatial state models are only one of many types of state model. Animal populations are heterogeneous in more than their spatial distribution, and we consider state models for all sources of heterogeneity.

9.2   State models

In Part II only plot sampling and distance sampling methods involved state models. Because these methods involve a single survey, occurring at what is effectively a single point in time, the associated state models need not have a temporal dimension. The mark-recapture and removal methods considered in Part II did not involve state models; those we consider below do. These methods involve multiple surveys, at different points in time, and this raises the question of whether and how animal-level variables might change between occasions. It introduces a temporal dimension into the state model.

Suppose we wanted to estimate the abundance of a population in which capture probability depends only on whether an animal is in woodland or grassland. Let $x_{si} = 0$ if animal i is in grassland at the time of survey s, and $x_{si} = 1$ if it is in woodland. If the N animals in a population behaved identically and independently, an appropriate state model on survey s would be

$$\pi(x_{s1}, \ldots, x_{sN})   =   \prod_{i=1}^{N}[1 - \phi_s]^{1-x_{si}}\phi_s^{x_{si}} \tag{9.1}$$

where ϕ_s is is the probability of being in woodland on occasion s.

As an animal moves between the two habitats from one survey occasion to another, it generates a "history" of locations, which we will call $\mathcal{H}$. In general there will be relevant animal-level variables other than location, and this history is a record of all the relevant animal-level variables (the $\underline{x}$s) of the animal on each survey occasion. We define $\mathcal{H}_{si}$ to be animal is history up to the start of occasion s:

$$\mathcal{H}_{si}   =   (\underline{x}_{1i}, \ldots, \underline{x}_{si}) \tag{9.2}$$

It is also convenient to have a symbol for the "history" of the population; we call it $\mathcal{H}_s$, which is a matrix whose rows are the histories of the individuals in the population ($\mathcal{H}_{s1}$ to $\mathcal{H}_{sN}$).

There are many ways in which the state model may change between surveys; we consider four below. Here, and throughout the book, we assume that animal-level variables are independent between animals. In some applications, this sort of model might be inadequate, but dealing with them is beyond the scope of this book. For simplicity, we formulate the state models in terms of a scalar x below, although in general it could involve a vector $\underline{x}$.

Static: Animals stay where they are (grassland or woodland) throughout the survey. Conceptually, the state model determines animal location only once, at the start of the survey, and from then on x_{si} is fixed ($x_{si} = x_{1i} = x_i$ for all s). The static state model for animal i can be written as follows[1]

$$\pi(\mathcal{H}_{si}) \quad = \quad \pi(x_{1i}) \qquad (9.3)$$

where x_{1i} is the first element of $\mathcal{H}_{si}$.

Independent dynamics: Animals move between habitats between surveys in such a way that the probability that an animal is in woodland (or grassland) is completely unrelated to where it was on previous surveys. Conceptually, the same state model applies anew for each survey (x_{si} is independent of x_{s^*i} for $s \neq s^*$). In this case, the state model for animal i can be written as

$$\pi(\mathcal{H}_{si}) \quad = \quad \prod_{s^*=1}^{s} \pi(x_{s^*i}) \qquad (9.4)$$

Markov dynamics: Animals' locations on each survey are influenced by their location on the most recent survey (only). Conceptually, the state model determines location anew for each survey, but the probability of an animal being in woodland or grassland depends on whether it was previously in woodland or grassland (x_{si} is random but depends on $x_{(s-1)i}$). We write the probability density of x_{si}, given $x_{(s-1)i}$, as $\pi(x_{si}|x_{(s-1)i})$. The state model for animal i can then be written as

$$\pi(\mathcal{H}_{si}) \quad = \quad \prod_{s^*=1}^{s} \pi(x_{s^*i}|x_{(s^*-1)i}) \qquad (9.5)$$

---

[1] Note that the probability density or mass functions on the left and right of Equation (9.3) are not the same, but for brevity we do not use different symbols for them; we rely on their arguments to distinguish them.

where $\pi(x_{1i}|x_{0i}) = \pi(x_{1i})$ is the state model at the start of the first survey.

Population-level drift: Animals' locations on each survey are determined independently of their previous positions, but there is "drift" in the distribution of animals, i.e. the distribution of the population changes as time progresses. Conceptually, the state model determines location anew for each survey, but the shape of the state model changes as time progresses – some of its parameters are time-dependent. To illustrate with the woodland-grassland scenario, we might have

$$\pi(\mathcal{H}_{si}) \quad = \quad \prod_{s^*=1}^{s} [1 - \phi(t_{s^*})]^{1-x_{si}} \phi(t_{s^*})^{x_{si}} \qquad (9.6)$$

where t_{s^*} is the time of survey occasion s^* and

$$\phi(t) \quad = \quad \frac{1}{1 + e^{\phi_o + \phi_t t}} \qquad (9.7)$$

for example, in which case, given enough time $(t \to \infty)$, animals stay in woodland or grassland (depending on the sign of ϕ_t) with no uncertainty.

In some applications, one of the above sorts of temporal dependence is the obvious or only reasonable one. In others, the choice may not be that clear, or an entirely different form of time-dependence might be most appropriate. In the sections that follow, we deal with at least one scenario in which each of the above forms might be appropriate. We do not consider other forms but hope that what we do gives an adequate basis for developing appropriate likelihood functions for them.

9.3   Observation models

In general, capture or detection probability may depend on some, all or none of the components of the animal-level variables, $\underline{x}_s$, as well as some, all or none of the components of the vector of survey-level variables, $\underline{l}_s$. It can also depend on capture history, $\underline{\omega}_s = (\omega_1, \ldots, \omega_{(s-1)})$ or the capture indicator, c_s (which is 1 if the animal has been captured before, and zero otherwise), or the individual animal identity if this is observable. Capture history is defined in terms of capture indicators for each occasion, as follows:

ω_{si} is the capture indicator for the ith animal on capture occasion s: $\omega_{si} = 1$ if animal i is captured on occasion s, otherwise it is zero.

$\underline{\omega}_{si}$ is the capture history of the ith animal immediately before capture
occasion s: $\underline{\omega}_{si} = (\omega_{1i}, \ldots, \omega_{(s-1)i})$.

The capture history of animals that have not been captured by the start
of occasion s is a vector of $(s-1)$ zeros, written $\underline{0}_s = (0, \ldots, 0)$.

There is no single best form of dependence of this probability on these
and other variables; we often use the logistic functional form, which is
a convenient general-purpose form for many applications. Whatever the
form, the detection/capture function has an associated unknown parameter
vector $\underline{\theta}$, which we usually do not make explicit in the likelihood functions,
for brevity.

10
Spatial/temporal models with certain detection

Key idea: "scale up" counts from covered region to survey region using a spatial/temporal model.

Number of surveys:	Single survey is adequate for spatial modelling; multiple surveys may be necessary for temporal modelling.
State model:	Spatial model for animal distribution in space and/or temporal model for animal distribution in time.
Observation model:	All animals in the covered region/period are detected with certainty.

10.1 Introduction

In this chapter, we consider model-based estimation of animal distribution when all animals in the covered region are detected. We look at estimation of

Likelihood functions and key notation:

Count data:

$$L(N;\underline{\phi}) \; = \; \left(\frac{N!}{(N-n)! \prod_{j=1}^{J} n_j!} \right) (1-\pi_c)^{N-n} \prod_{j=1}^{J} \pi_j$$

Individual location/time data:

$$L(N;\underline{\phi}) \; = \; \binom{N}{n} (1-\pi_c)^{N-n} \prod_{i=1}^{n} \pi(\underline{x}_i)$$

N: number of animals in the population (abundance).

J: number of covered units.

n_j: number of animals counted in covered unit j.

n: $= \sum_{j=1}^{J} n_j$; total number of animals counted.

$\underline{x}_i$:   time (migration survey) or location (plot survey) at which the ith detected animal is detected.

$\pi(\underline{x})$:   probability density of $\underline{x}$, i.e. the temporal/spatial state model.

π_j: probability that a given animal is in the jth covered unit.

π_c: $= \sum_{j=1}^{J} \pi_j$; probability that a given animal is in the covered region.

$\underline{\phi}$:   the parameters of the temporal or spatial state model.

- spatial state models from plot surveys,

- temporal state models from migration counts, and

- spatio-temporal state models from a series of plot surveys.

Migration counts and plot surveys are conceptually very similar. While a plot survey comprises a sample of counts from a number of plots at a single time, a migration count survey comprises a sample of counts at a number of times. The periods during which migration counts are made are called watch periods; they are analogous to the sampled plots in a plot survey. In the same way that animals in the sampled plots are detected, in a simple migration count all animals passing during a watch period are counted (by assumption). The migration survey problem involves estimation of the number of animals passing outside of surveyed periods; the plot survey problem involves estimation of the number of animals located outside of surveyed plots.

For migration surveys, the state model is temporal, while it is spatial for plot surveys.

When dealing with plot survey methods in Chapter 4, we assumed that animals were uniformly distributed in space, but noted that this assumption would often be violated. The assumption of uniform temporal distribution in a migration is usually even less plausible, but we often know that the temporal distribution of migrating animals has a fairly simple form. It is one-dimensional (time being the dimension) and often smooth; the number of animals passing a watch point often builds up smoothly through time from zero to some maximum, and then falls off smoothly to zero again by the end of the migration season. In the case of plot sampling, the spatial distribution of animals in the survey region is two-dimensional and rarely has such a simple form. Migration counts provide a context in which a relatively simple but nevertheless plausible non-uniform temporal state model can be useful in estimation. We deal with migration counts first and then go on to the slightly more complicated problem of estimating abundance from plot surveys with two-dimensional spatial models. Finally, we look at an example which involves both space and time – estimating the abundance of a closed but transitory population from multiple surveys, using a spatio-temporal state model.

10.2   Temporal modelling: migration counts

Migration counts to date have been used exclusively for estimating the size of populations of whale that migrate annually past coastal watch points. They are potentially relevant to any population of migratory animals for which it is possible to survey an area through which the whole population must pass.

The California gray whale (*Eschrichtius robustus*) population is surveyed from Monterey by this method as it migrates between its feeding grounds in the Bering and Chukchi Seas and its calving grounds in Baja California. At Monterey almost the entire population passes within a few kilometres of the coast (Reilly *et al.*, 1980, 1983). Counts of whales passing lookout points are made regularly during daylight hours throughout the migration. From the observed counts we want to estimate the total number of animals that passed Monterey over the whole migration period.

10.2.1   Estimation by design

If we can assume that the watch periods are a random sample of periods from throughout the migration, then design-based methods identical to those used for plot sampling (Chapter 4) can be used. Conceptually, the population can be thought of as a one-dimensional "queue" of animals,

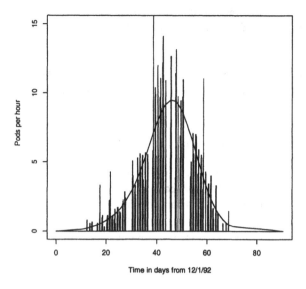

Figure 10.1. Observed counts in watch periods (the bars) and fitted temporal state model (the curve) from the 1992 survey of the Californian gray whale migration past Monterey.

which is sampled by randomly placing plots along the length of the queue. All the results of Section 4.3 then follow. The problem with this approach is that random designs are impractical. Counts cannot usually be conducted at night or in bad weather, and, to improve efficiency, more counts tend to be made at the height of the migration than early or late. While this latter problem may be partially overcome by stratification of the migration period, users of migration counts prefer model-based analysis, for which the assumption of a random sample of possible watch periods is not required.

10.2.2   Maximum likelihood estimation

To adopt the model-based methods of Chapter 4, we would be forced to assume that animals pass the watch point independently at a uniform rate throughout the migration period. This assumption is untenable (Figure 10.1).

The number of animals passing per unit time changes through the migration period. To model this, we specify a temporal state model for the migration by defining a migration rate of $\lambda(t)$ animals per unit time at time t, with $t = 0$ corresponding to the onset of migration, and $t = T$ to the end of the migration. The smooth curve in Figure 10.1 is proportional to the migration rate. In this example, time is measured in days and $\lambda(t)$ starts at zero at day zero, builds up to a maximum around day 45, and then drops off to zero again by about day 90. In their analyses, Breiwick *et al.* (unpublished) and Buckland and Breiwick (in press) modelled $\lambda(t)$ using a sum of Hermite polynomials, but any suitable form could be used for this function. In any event, $\lambda(t)$ is the basis of the temporal state model and has a vector of unknown parameters, $\underline{\phi}$, associated with it. We estimate these parameters by maximum likelihood. The technical section below gives the details.

"Binned" likelihood for counts

Technical section ↓

To form a likelihood, we need to obtain an expression for π_j, the probability that a given animal passes the watch point during watch period j. The probability that an animal passes the watch point in the small time interval $(t, t+dt)$ is $\lambda(t) \times dt/(\int_0^T \lambda(t)\, dt)$ and the corresponding state model for the time at which the animal passes the point is

$$\pi(t) = \frac{\lambda(t)}{\int_0^T \lambda(t)\, dt} \tag{10.1}$$

Note that if $\lambda(t)$ is a probability density function (which it need not be), then $\pi(t) = \lambda(t)$. The probability that the animal passes the watch point in watch period j which starts at time t_{s_j} and finishes at time t_{f_j}, is therefore

$$\pi_j = \int_{t_{s_j}}^{t_{f_j}} \pi(t)\, dt \tag{10.2}$$

The probability π_j depends on parameters $\underline{\phi}$, and it is sometimes useful to make this explicit, i.e. to write it as $\pi_j(\underline{\phi})$, but for brevity we do not usually write it this way. Because a given animal can pass during one period only, the probability that an animal passes during one of the J watch periods (i.e. the "covered period") is just the sum of the π_js:

$$\pi_c = \pi_c(\underline{\phi}) = \sum_{j=1}^{J} \pi_j(\underline{\phi}) \tag{10.3}$$

If we assume that the time any particular animal passes is independent of the times other animals pass, and we think of the event "pass in a watch period" as a "success", then the likelihood for N and $\underline{\phi}$ is the following multinomial (see Exercise 10.1)

$$L_b(N;\underline{\phi}) = \left(\frac{N!}{(N-n)!\prod_{j=1}^{J} n_j!}\right)(1-\pi_c)^{N-n}\prod_{j=1}^{J}\pi_j^{n_j}$$
(10.4)

The likelihood is determined by the state model for migration rate, $\pi(t)$. As with plot sampling, for the basic method, there is no randomness in the observation model because detection is certain within watch periods. The likelihood Equation (10.4) may be maximized with respect to N and $\underline{\phi}$ using standard numerical procedures.

Another approach, used in practice, is to estimate $\underline{\phi}$ by maximizing the conditional likelihood, given n (see Exercise 10.1(a)) with respect to $\underline{\phi}$. Having obtained an estimate $\tilde{\underline{\phi}}$ of $\underline{\phi}$, π_c may be estimated by $\tilde{\pi}_c = \pi_c(\tilde{\underline{\phi}})$. N can then be estimated by maximum likelihood, given $\tilde{\pi}_c$ (see Exercises 10.1(b) and (c)). This gives the conditional MLE $\tilde{N} = n/\tilde{\pi}_c$. Sanathanan (1972) showed that asymptotic distributions of the conditional MLEs $\tilde{N}$ and $\tilde{\underline{\phi}}$ are identical to those of the unconditional MLEs $\hat{N}$ and $\hat{\underline{\phi}}$, obtained by maximizing the full likelihood Equation (10.4) with respect to N and $\underline{\phi}$.

↑
Technical
section

Interval estimation

Profile likelihood methods or the parametric bootstrap can be used to estimate confidence intervals, by generating resamples from the fitted multinomial distribution. The nonparametric bootstrap, although implemented differently, is equivalent to the parametric bootstrap in this situation. It may be applied by listing all animals and resampling them with replacement. Thus n_j animals are listed as counted in watch period j, $j = 1, ..., J$, and a further $\hat{N} - n$ animals are listed as not counted (where $n = \sum n_j$). $\hat{N}$ animals are then selected with replacement from this list for each resample. Analysis now proceeds as described previously for the bootstrap.

10.2.3 Inference with "Unbinned" data

Estimation from migration count surveys generally involves using the counts in each watch period rather than the observed times each animal passed the watch points. The watch periods are usually so short in comparison to the duration of the migration that little would be gained by using the individual times. It is, however, quite easy to extend the likelihood of Equation (10.4) for the case in which individual times are recorded. Details are contained in the technical section below. Profile likelihood, parametric bootstrap, or nonparametric bootstrap methods can be used here too for estimation of confidence intervals.

"Unbinned" likelihood for times

Technical section
↓

To develop a likelihood function for use with individual detection times, we need the conditional probability density of passing time, given that the animal was detected. Let the time that the ith detected animal in the jth watch period passes the observer be t_{ij}. The unconditional pdf of t_{ij} is $\pi(t_{ij})$; using Bayes' theorem, and the assumption that animals pass independently of one another, the joint probability density of the times of passing is

$$L_u(\underline{\phi} \mid \underline{n}) \;=\; \prod_{j=1}^{J}\prod_{i=1}^{n_j} \frac{\pi(t_{ij})}{\pi_j} \qquad (10.5)$$

Multiplying Equation (10.4) by this, we get

$$L_u(N;\underline{\phi}) \;\propto\; \binom{N}{n}(1-\pi_c)^{N-n} \prod_{k=1}^{n} \pi(t_k) \qquad (10.6)$$

where t_k is the time that the kth detection passes.

↑
Technical section

10.3   Spatial modelling: plot sampling

The estimation of abundance by modelling the spatial distribution of animals with plot sample data is the same as the estimation of abundance from a simple migration survey, except that:

(1) with a migration survey the survey region is one-dimensional and the dimension is time, while in a plot survey, the survey region is two-dimensional and the dimensions are spatial;

(2) with a migration survey the state model usually has a relatively simple shape, whose key features are well known before the survey, while the spatial distribution of animals in a plot survey can be very complicated with unknown key features.

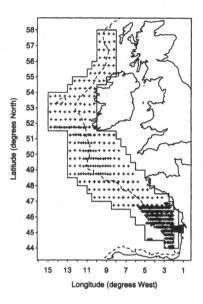

Figure 10.2. Survey region and sample locations for the 1992 survey of the western mackerel stock. The dashed line is the 200 m depth contour.

These differences (and the second in particular) probably account for the fact that few examples of modelling the spatial distribution of animals from plot surveys seem to have appeared in the literature. Nevertheless, with a suitably flexible form for the state model describing animals' spatial distribution, estimation of abundance with a state model is quite feasible and may greatly improve estimation precision.

A real example of the utility of this approach is the estimation of the abundance of mackerel off the west coasts of France, Britain and Ireland from egg survey data. The stock is assessed using the "Daily Egg Production Method" (DEPM: Gunderson, 1993; Hunter and Lo, 1993). This involves estimating egg abundance on a single day (ideally) from a shipboard survey of eggs, and estimating the mean number of eggs produced per fish per day from a simultaneous survey of fish. The method is an example of an indirect assessment method, in which fish abundance is estimated from an estimate of egg abundance. In this chapter we focus on the egg survey alone.

The survey region and location of net samples of eggs for the 1992 survey is shown in Figure 10.2.

Each sample location (net haul) can be thought of as a plot sample unit, and the nets are such that all eggs within the plot are observed. The density of mackerel eggs in the survey region is known to be very non-uniform. While it varies in detail from survey to survey, it has some persistent features. There is a considerable change in density from zero in the east, to a peak usually near the 200 m depth contour, back down to very low density on the western border of the survey region. There is also considerable systematic change in density from south to north: density is low in the south, highest at middle latitudes, and very low again in the north. The water around the "porcupine bank", a shallow water region centred on $14°W$, $53.5°N$, is usually also a region of relatively high egg density.

The density surface is a function of location, depth, and possibly other variables, but we start by considering it to be a function of location alone. We refer to the density surface as $\lambda(\underline{x})$, where $\underline{x} = (u, v)$ and u and v are longitude and latitude. Because the shape of the surface is not very predictable, a flexible nonparametric functional form must be used for $\lambda(\underline{x})$. In the example described here, $\lambda(\underline{x})$ was modelled using smoothing splines (see Hastie and Tibshirani, 1990).

With this form, the state model for the location of any egg is

$$\pi(\underline{x}) = \frac{\lambda(\underline{x})}{\int \int \lambda(\underline{x}) \, du \, dv} \tag{10.7}$$

where integration is over the whole survey area. The probability that an egg is in the jth haul is modelled as

$$\pi_j = \int \int \pi(\underline{x}) \, du \, dv \tag{10.8}$$

where integration is over the range of u and v covered by the jth haul. (In this application, this area is such a tiny fraction of the survey region that nothing is lost by approximating this probability by $\lambda(\underline{x}_j)a_j$, where a_j is the effective area of the jth net and $\underline{x}_j$ is the centre of the haul.) A total of n_j mackerel eggs is counted in the jth haul.

In this application, distance from the 200 m depth contour (D200) and bottom depth (BDp) were found to be important explanatory variables and were included in the model. The model above is easily extended to include these variables by defining $\underline{x} = (u, v, D200, BDp)$. Note that D200 and BDp are known everywhere in the survey region.

The nonparametric form for $\lambda(\underline{x})$ makes estimation by maximization of the likelihood Equation (10.4) infeasible, and in this case estimation was performed using Generalized Additive Models (GAMs; Hastie and Tibshirani, 1990).[1]

---

[1]The model used in that paper is slightly different to the model we discussed above; the main difference being that a Poisson density was assumed for n, rather than the binomial density underlying Equation (10.4) (see Exercise 10.1(a)). One can think of the

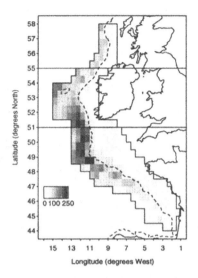

Figure 10.3. Estimated density surface from the analysis of the 1992 survey of the western mackerel stock. The dashed line is the 200 m depth contour; the two horizontal lines are the borders between the strata used in the design-based estimation of egg production from these data.

The estimated egg density surface is shown in Figure 10.3; the estimated effect of latitude, longitude, distance from the 200 m contour, and bottom depth, are shown in Figure 10.4.

Interval estimation

Variance and confidence interval estimation can proceed as for migration counts. A bootstrap method that involves sampling plots with replacement will have poorer coverage of the survey region than the original survey (because in any bootstrap sample some plots will be omitted while others will be chosen more than once). A parametric bootstrap does not suffer from this difficulty, but the parametric model does need to be appropriate. In the mackerel example, egg counts were found to be substantially more variable than the parametric (Poisson) model predicted; a method due to

difference between use of a Poisson model and binomial models either as a philosophical one, or as an approximation. With the former view, we estimate the underlying density of a notional hyper-population from which the actual population of N eggs was obtained, while with a binomial model, we estimate the abundance of the actual population. With the latter view, the Poisson is a convenient approximation for the binomial. Details of the full analysis can be found in Borchers *et al.* (1997), Bowman and Azzalini (1997) also discuss estimation of a density surface for these data.

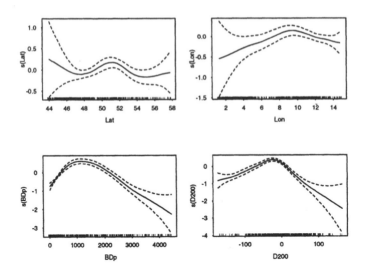

Figure 10.4. The estimated effects of latitude, longitude, bottom depth (BDp) and distance from 200 m contour (D200) on mackerel egg density. The solid lines represent the estimated effect; the dashed lines are pointwise 95% confidence limits and the ticks along the bottom axis indicate the locations of sample data.

Bravington (1994) was used to incorporate this "overdispersion" in the parametric bootstrap.

Summary

Because a large component of the variance in egg numbers through the survey region in this example is systematic spatial trend in density, modelling the distribution of eggs resulted in a large increase in the precision of the egg abundance estimate over that obtained using conventional (stratified) design-based estimation methods.

Unlike design-based methods, it also provides a smooth density surface and information about which variables are associated with high animal densities – something that may be of interest in itself. We anticipate that similar methods will see increasing use in future.

10.4   Spatio-temporal modelling

In some situations it makes sense to model both the spatial distribution of a nominally closed population, and how it changes with time. The estimation of the abundance of mackerel off the west coasts of France, Britain and Ireland from egg survey data, again provides a real example. The state model in this case is one with what we called "population level drift" in

Chapter 9. The "Annual Egg Production Method" (AEPM; see Gunderson, 1993) is routinely used to assess this stock. It involves estimating the total egg production in a spawning season, not just egg abundance on a day. Data are gathered from a number of egg surveys spread through the spawning season, each a smaller version of the DEPM survey described in the preceding section.[2]

Figure 10.5 shows the location of samples in the 1995 AEPM survey.

We briefly summarize results from estimation of mackerel egg production from the 1995 survey. Detailed methods and results are given in Augustin et al. (1998b). Density was modelled as a function of four variables: distance from the 200 m depth contour, distance north along this contour, bottom depth, and date. Again the form of dependence on spatial and temporal variables is not well known and may be complicated, so that $\lambda(\underline{x})$ was modelled using smoothing splines, and estimation was performed using GAM methods. Details can be found in Augustin et al. (1998b); the key point here is that modelling the egg distribution in space and time again resulted in improved estimation precision. Plots of the estimated distribution of mackerel eggs from the 1995 survey are shown in Figure 10.6.

Figure 10.7 shows the estimated egg production through time in 1995. It is analogous to the smooth curve representing whale migration in Figure 10.1.

10.5   Other methods

In the above examples, we have not modelled temporal or spatial autocorrelation. There is a good reason for this. Generally, wildlife population managers want to quantify the size of a population, and perhaps its spatial distribution, at the time of the survey(s). There is no intrinsic interest in whether animals are clustered in space or time through true autocorrelation (animals like to be with, or near, other animals) or whether the clustering occurs due to (potentially measurable) covariates, such as habitat in a spatial survey or weather conditions in a temporal survey. If we can identify a model that assumes independence and that also fits the data adequately (through the use of proxy covariates if necessary), then we can draw valid inference on population size. If part or all of the animal distribution came about as a result of correlation rather than trend, but we model it as trend,

---

[2]With a bit of licence, you could think of the AEPM method as combining a kind of migration survey with plot surveys. Each complete survey in the spawning period corresponds to a notional "watch point", while the egg density in the survey region increases and decreases smoothly with time in much the same way that animal density in the survey region would in a proper migration survey. The survey has a spatial component not usually present in migration surveys, because of the spatial trend in egg density in the survey region.

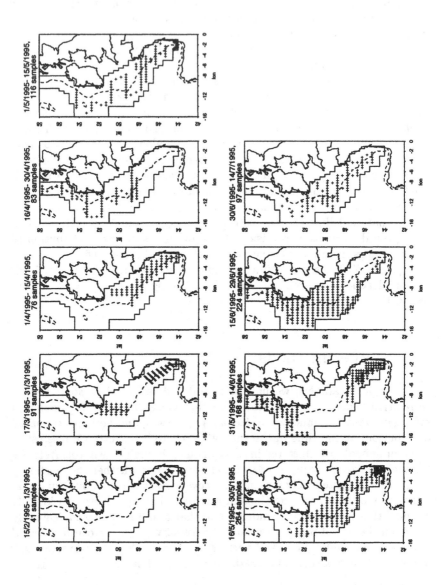

Figure 10.5. Survey region and sample locations for the 1995 survey of the western mackerel stock. The dashed line is the 200 m depth contour.

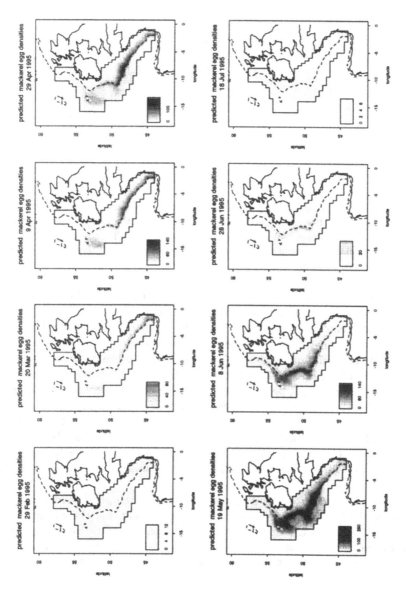

Figure 10.6. Estimated egg density surface from the analysis of the 1995 survey of the western mackerel stock. The dashed line is the 200 m depth contour.

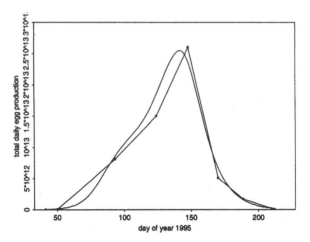

Figure 10.7. Estimated daily mackerel egg production through time in 1995. The smooth curve is that estimated using generalized additive models; the piecewise linear curve is obtained by treating the survey data as if they came from four points in time (at the dots joining the straight lines) and interpolating linearly between them in time.

our state model can be thought of as a prediction surface for the realized distribution. If instead we use geostatistical methods (Cressie, 1998) or autocovariate models (Augustin *et al.*, 1996) to model autocorrelation, then conceptually, we are drawing inference on a super-population, from which our actual population, with its observed clusters of distribution, is drawn (Augustin *et al.*, 1998a). The practical implication is that our variance estimates are then larger, because they include a component corresponding to the stochastic process of drawing our population from the super-population. By not modelling the autocorrelation, we make predictions about the actual realized population, not the underlying super-population. It is usually the former that is of interest to the wildlife manager. Nevertheless, if we want to model autocorrelation, then the methods outlined above can be extended.

In most applications, if a temporal or spatio-temporal model is to be used to fit data from a series of surveys, it is likely that the wildlife manager will want to use open population models, to allow for birth, death and movement. This book is primarily concerned with closed populations, although we do touch on open population models in Chapter 13.

10.6 Summary

The methods of this chapter do not involve observation models – all animals in the covered region are detected. They can be thought of as extensions of plot sampling methods, the main extension being the use of non-uniform state models. The resulting methods are useful for modelling animal distributions in both time and space. As well as improving the precision of abundance estimates in cases where there is considerable variation in animal density in time and/or space, they provide estimates of the spatial and/or temporal distributions of animals in the survey region. These are often of interest in their own right.

While the methods are not conceptually much more challenging than plot survey methods, they are more complicated in their application. Usually relatively little is known about the spatial and/or temporal distributions, except that they should be smooth and possibly that densities should be low or zero at the extremes of the survey region or survey times. Flexible forms of state model are required as a result. Generalized linear or generalized additive modelling methods provide powerful tools for estimating these flexible state models.

An advantage of our emphasis on unified likelihood methods is that it provides a framework for models that combine different methodologies. This should be apparent from the AEPM survey example above. Other useful extensions are also possible. In many applications, some animals in the covered region go undetected. In migration surveys, for example, animals may be missed even though they pass in a watch period. A solution is to use two observers, or observer teams, who independently record passing animals. In this case, an observation model is required and the resulting methods can be thought of as a combination of simple migration count survey methods and mark-recapture methods. They usually involve modelling heterogeneity (changes in detection/capture probability due to animal-level variables). We discuss this approach in Chapter 12, after we cover ways of dealing with heterogeneity in Chapter 11.

10.7   Exercises

Exercise 10.1   Consider a migration count of a population, with J watch periods. Suppose that animals pass the watch point independently of one another, and the probability that an animal passes during one of these watch periods is $\pi_c \equiv \pi_c(\underline{\phi})$, where $\underline{\phi}$ is an unknown parameter.

(a) Obtain the likelihood function for the number of animals passing in total (N) if π_c is known and n animals are counted in all.

(b) If $\pi_j \equiv \pi_j(\underline{\theta})$ is the probability that an animal passes in watch period j, write down an expression for the probability that an animal passes in period j, given that it passed in one of the watch periods ($j = 1, \ldots, J$).

(c) Suppose n_j animals are counted in period j $(j = 1, \ldots, J)$. Using (b), obtain the likelihood for $\underline{\phi}$, given n_j $(j = 1, \ldots, J)$.

(d) Hence show that the likelihood for N and $\underline{\phi}$ is given by Equation (10.4).

11

Dealing with heterogeneity

Key idea: use state model to describe the characteristics of unobserved animals probabilistically.

Number of surveys:	One or multiple; depends on method.
State model:	Probability model for animal-level variables that affect detection/capture probability. Animal-level variables may change with time.
Observation model:	Various; depends on method.

11.1 Introduction

We use the term "heterogeneity" to refer to differences in capture or detection probability at the level of individual animals or groups of animals. For our purposes, a heterogeneous population is one in which individuals or groups in the covered region are not equally detectable even when exactly the same survey effort is applied to them. (Note that heterogeneity in this sense is a product of both the physical properties of the population and the survey method: the same population might be heterogeneous with one survey method but not with another.)

Of the methods dealt with in Part II, only distance sampling methods accommodate heterogeneity. The change-in-ratio method involves animal-level differences, but assumes they do not affect capture probability. Distance sampling methods are based on changes in detection probability due to the animal-level variable distance. They accommodate and model it by incorporating a probability model (a state model) for the distribution of distance in the population. This is given by Equation (7.1) for line transect surveys, and Equation (7.30) for point transect surveys.

The key role of the state model when there is heterogeneity is apparent from line transect surveys. The only reason that we are able to interpret the drop-off in detection frequency as distance from the line increases (as in Figure 7.5, for example) is that the state model implies the same number of animals (detected or not) at all distances, on average. Without the state model, we would have no way of distinguishing a drop-off in frequency caused by fewer animals further from the line, from a drop-off due to lower detection probability further from the line.

Without heterogeneity, we know that the missed animals were as likely to be detected as the detected animals. When there is heterogeneity, we don't know how difficult missed animals were to detect. If they were all very difficult to detect, we would expect to have missed many and expect the population to be larger. If they were all very easy to detect, we would expect to have missed few and expect the population to be little bigger than the observed population. A state model that incorporates the source of heterogeneity (distance in line transect surveys, for example) gives us the means to make inferences about the detectability, and hence the number, of animals we missed.

By definition, plot sampling does not involve heterogeneity because all animals in the covered region are detected with certainty. Distance sampling involves heterogeneity and a state model for heterogeneity. The simple removal methods and mark-recapture methods of Part II do not accommodate heterogeneity. We now develop these methods for heterogeneous populations, by allowing capture/detection probability to depend on animal-level variables and including appropriate state models in the likelihoods.

11.1.1   Combining state and observation models

Although we do not always construct them in this way explicitly, a useful way to think of the construction of the likelihood models of this section is as follows.

(1) Specify a state model for the animal-level variables. This, considered as a function of the state model parameters N and ϕ, is the likelihood for N and $\underline{\phi}$, given the animal-level variables; call it $L(N, \underline{\phi})$.

(2) Specify an observation model for what is observed on each of the S surveys, given the state model parameters and the animal-level variables; call it $L(\underline{\theta}|N,\underline{\phi})$, where $\underline{\theta}$ is the observation model parameters. Now $L(N,\underline{\phi}) \times L(\underline{\theta}|N,\underline{\phi})$ is the likelihood for N, $\underline{\phi}$ and $\underline{\theta}$, given the observed data and the animal-level variables.

(3) Not all the animal-level variables are observed, so we can't base inference on $L(N,\underline{\phi}) \times L(\underline{\theta}|N,\underline{\phi})$. In order to get the likelihood, given what was observed, we integrate out the unobserved animal-level variables to get a marginal likelihood, given what was observed.

In the case of multiple survey methods, it is sometimes convenient to construct the likelihood sequentially; we construct the likelihood for survey occasion s, given what has been observed on preceding occasions. We do so for mark-recapture methods below.

11.2   Distance sampling with covariates

In this section we deal with ways of incorporating variables[1] into distance sampling likelihoods. We consider two sorts of variable: animal-level variables and survey-level variables. We develop a likelihood for the former using the animal-level variable group size (in which case groups are treated as the "animals") but the likelihoods are applicable to animal-level variables generally. When we deal with survey-level variables, we consider only the case in which these are known throughout the survey region. We develop a likelihood using observer as the survey-level variable. The development is applicable for any survey-level variable which is constant for the period during which an animal is available for detection. If we have survey-level variables that can change substantially over the period an animal is available, a slightly different treatment is required. In practice, survey-level variables seldom change sufficiently fast to make this a real problem.

Because they are not known for every animal in the population, incorporating animal-level variables requires incorporation of a state model for the distribution of these variables in the population. This is not the case for the survey-level variables we consider because they are known throughout the survey region.

Recall that with sufficiently flexible detection functions, distance sampling estimators are pooling robust; i.e. neglecting animal-level or survey-level variables does not introduce bias. So why might we want to incorporate

---

[1] The distance sampling literature tends to use the term "covariate" where we use the term "variable". We use the two interchangeably, preferring "variable", for consistency with the rest of this book.

variables in distance sampling methods when we have pooling robustness? Reasons include:

(1) The detection functions at different levels of the variables might be sufficiently different that without very large sample size, pooling robustness fails.

(2) There may be interest in the variables themselves. An animal-level example is group size; the distribution of group sizes in the population or the mean group size might be of interest in its own right. A survey-level example is observer; there may be interest in how good each of the observers on a survey are.

(3) It may improve estimation precision if detection functions at different levels of variables are very different.

In the sections that follow, we develop a full likelihood function that incorporates animal-level variables and then extend it to incorporate survey-level variables. We also develop a likelihood function that includes both sorts of variable, conditional on the number of animals observed and the values of variables other than x. The MLEs of the detection function from the conditional likelihood can be used with a Horvitz–Thompson-like estimator for partly design-based estimation of abundance.

11.2.1 Full likelihood: animal-level variables

Technica
section
↓

The distance sampling likelihood of Equation (7.10) can readily be extended to accommodate groups. Denote the probability distribution of group sizes in the population by $\pi_z(z; \underline{\phi})$, $z = 1, 2, ...$, where $\underline{\phi}$ are the parameters of the group size distribution.

Sampling without size bias

When detection probability does not depend on size, the contribution of group sizes to the likelihood is

$$L_z(\underline{\phi}) \quad = \quad \prod_{z=1}^{\infty} \pi_z(z; \underline{\phi})^{n_z} \qquad (11.1)$$

where n_z is the number of groups of size z detected ($\sum_z n_z = n$). For brevity we usually write $\pi_z(z; \underline{\phi})$ as $\pi_z(z)$; we write the probability distribution of distance $\pi_x(x)$. A model such as the negative binomial can now be specified for $\pi_z(z)$, and fitted by maximizing this likelihood, to give $\hat{\pi}_z(z) \equiv \pi_z(z; \hat{\underline{\phi}})$. This yields the MLE of mean size of groups, $\hat{E}[z] = \sum_{z=1}^{\infty} z\hat{\pi}_z(z)$. The MLE $\hat{N}$ of Section 7.2.2 now estimates the number of groups in the population, so population abundance is estimated by multiplying this by $\hat{E}[z]$.

In practice, it is simpler, and consistent with a design-based estimation philosophy, to use $\bar{z}$ as an unbiased estimate of $E[z]$. This estimate is also likely to be more robust, given the difficulty in specifying a suitable model for $\pi_z(z)$.

Size-biased sampling

If probability of detection is affected by group size, then the full likelihood can be written

$$L(N, \underline{\theta}, \underline{\phi}) = \binom{N}{n} (1 - \pi_c E[p])^{N-n} \prod_{i=1}^{n} \pi_c p(\underline{x}_i) \pi(\underline{x}_i)$$

(11.2)

where $\underline{x}_i = (x_i, z_i)^T$ is the vector of animal-level variables associated with animal i, $\pi(\underline{x}) = \pi_x(x)\pi_z(z_i)$ is the probability distribution of $\underline{x}$ (assuming distance, x, to be independent of group size, z), and $E[p]$ is the expected value of the detection function $p(\underline{x})$ with respect to $\underline{x}$:

$$E[p] = \int_0^w \sum_{z=1}^{\infty} p(\underline{x})\pi(\underline{x})\, dx = \int_0^w \sum_{z=1}^{\infty} p(x, z)\pi_z(z)\pi_x(x)\, dx$$

(11.3)

Thus we need to specify a model for the probability density of $\underline{x}$ in the population, $\pi(\underline{x})$, and a model for detection probability as a function of $\underline{x}$. A reasonable way of including animal-level variables into the detection function is to have them affect only the scale parameter (see Drummer and McDonald, 1987; Ramsey *et al.*, 1987; Marques and Buckland, 2003). Having done this, we maximize the likelihood with respect to the parameters of the detection function, $\underline{\theta}$, and the parameters $\underline{\phi}$ of the pdf $\pi(\underline{x})$. Often it will be appropriate to assume that distance is independent of other components of $\underline{x}$, as we have done above, but the likelihood Equation (11.2) is applicable with any appropriate $\pi(\underline{x})$.

Note that if $p(\underline{x}) = 1$ for animals in the covered region and zero outside it, and we define $\pi^*(\underline{x}) \equiv \pi_c \pi(\underline{x})$, then the likelihood Equation (11.2) is identical in form to the likelihood Equation (10.6) for migration counts.

↑
Technical
section

11.2.2   Full likelihood: survey-level variables

Including survey-level variables does not involve much more than using a detection function model that incorporates the variables appropriately. If the survey involved say K observers, with only one on duty in any one place at any one time, observer index (call it $l \in \{1, 2, \ldots, K\}$) would be an appropriate survey-level variable. Because we know which observer was on duty at any time and location, we can associate a unique observer with every detected animal and with every section of transect line (line transect sampling), or point (point transect sampling).

Technical
section
↓

In addition to its perpendicular distance, x_i, the ith detected animal has the survey-level variable l_i associated with it. As an example, consider using a half-normal detection function in which observer index affected the scale parameter. In this case

$$p(x_i, l_i) = \exp\left\{\frac{-x^2}{2\,\sigma(l_i)^2}\right\} \qquad (11.4)$$

where $\sigma(l_i)$ is some suitable function of observer index l_i; for example:

$$\sigma(l_i) = \exp\{\theta_0 + \theta_{l_i}\} \qquad (11.5)$$

where θ_{l_i} is the detection function parameter associated with observer l_i.

The term $E[p]$ in the likelihood is the mean value of the detection function in the covered region. If observer l was on duty for a fraction π_l of the total search effort (distance in line transect surveys, points in point transect surveys), then

$$E[p] = \int_0^w \sum_{l=1}^K p(x, l)\pi_l \pi_x(x)\, dx \qquad (11.6)$$

Note the similarity with Equation (11.3). A notable difference is that we do not need to specify a model for π_l the way we did for $\pi_z(z)$ – because the proportion of the survey with the survey-level variable equal to l is known, whereas the proportion of the population with the animal-level variable group size equal to z is unknown.

↑
Technical
section

11.2.3   Conditional likelihood: estimation partly by design

Recall that conventional distance estimation methods use the conditional likelihood, given detection, to estimate the parameters of the detection function (see Section 7.2.2). In a similar way, one can estimate the parameters of the detection function when it depends on distance x and other variables, from the conditional likelihood, given detection and the values of variables other than x.

Technical
section
↓

Notation: The vector $\underline{x}$ is the vector of all relevant animal-level variables; x is perpendicular distance; z is size; let $\underline{y}$ be a vector of all relevant animal-level variables aside from x and z. So[2] $\underline{x} = (x, z, \underline{y}^T)^T$.

The conditional density of x_i, given z_i, $\underline{y}_i$ and l_i is

$$f(x_i \mid z_i, l_i) = \frac{p(\underline{x}_i, l_i)\pi(x_i \mid z_i, \underline{y}_i)}{\int_0^w p(\underline{x}_i, l_i)\pi(x_i \mid z_i, \underline{y}_i)\, dx} \qquad (11.7)$$

from which we get the conditional likelihood

$$L(\underline{\theta}) = \prod_{i=1}^n \frac{p(\underline{x}_i, l_i)\pi(x_i \mid z_i, \underline{y}_i)}{\int_0^w p(\underline{x}_i, l_i)\pi(x_i \mid z_i, \underline{y}_i)\, dx} \qquad (11.8)$$

and if x, z and $\underline{y}_i$ are independent, $\pi(x_i \mid z_i, \underline{y}_i) = \pi_x(x_i)$, so that

$$L(\underline{\theta}) = \prod_{i=1}^n \frac{p(\underline{x}_i, l_i)\pi(x_i)}{\int_0^w p(\underline{x}_i, l_i)\pi(x_i)\, dx} \qquad (11.9)$$

Given an estimate $\hat{\underline{\theta}}$ from this likelihood, group abundance and individual abundance can be estimated using Horvitz–Thompson-like estimators:

---

[2]Don't confuse $\underline{x}$ and x in this section: the former is the vector of all animal-level variables, the latter is the animal-level variable distance.

$$\hat{N}_{grp} = \sum_{i=1}^{n} \frac{1}{\frac{a}{A}\hat{p}(\underline{x}_i, l_i)} \quad \text{and} \quad \hat{N}_{ind} = \sum_{i=1}^{n} \frac{z_i}{\frac{a}{A}\hat{p}(\underline{x}_i, l_i)}$$

$$(11.10)$$

where $\hat{p}(\underline{x}_i, l_i)$ is $p(\underline{x}_i, l_i)$ evaluated at $\underline{\hat{\theta}}$. Alternatively, the following estimators use our knowledge of $\pi_x(x)$ in the covered region and will be less variable:

$$\hat{N}_{grp} = \sum_{i=1}^{n} \frac{1}{\frac{a}{A}\hat{E}_x[p(\underline{x}_i, l_i)]} \quad \text{and} \quad \hat{N}_{ind} = \sum_{i=1}^{n} \frac{z_i}{\frac{a}{A}\hat{E}_x[p(\underline{x}_i, l_i)]}$$

$$(11.11)$$

where

$$\hat{E}_x[p(\underline{x}_i, l_i)] = \int_0^w p(x, z_i, \underline{y}_i, l_i)\pi_x(x)\, dx \quad (11.12)$$

↑
Technica
section

11.3   Mark-recapture

Animal populations are almost always heterogeneous and mark-recapture estimators are biased when heterogeneity is neglected (Chapter 6).

Variables causing heterogeneity might be observable (when capture probability depends on sex or age, for example, this might be the case), or not (e.g., unobservable heterogeneity in animal behaviour). Different methods are required for the two cases.

Observed heterogeneity There are two broad methods for dealing with heterogeneity when the variables causing it can be observed.

Animal-level stratification The simplest method is to stratify the analysis by these variables. For example, if males were more (or less) catchable than females, we could stratify by sex ($x = 0$ for females, $x = 1$ for males) and estimate the abundance of females $N(0)$ and the abundance of males $N(1)$. With this approach, no state model is necessary – the numbers of animals of each sex are treated as parameters rather than realizations of a state model determining sex. This is not necessarily the same as treating the mark-recapture experiment as two separate experiments (one for each sex), because the capture function for females might share parameters with that for males. Only if the two capture functions are estimated completely separately does this approach become equivalent to two separate experiments.

Use of a state model In some situations in which the variables causing heterogeneity can be observed, animal-level stratification is not desirable or feasible. Consider the case in which capture consists of detection by an observer and the variable causing heterogeneity is distance x of the animal from the observer. Here x is a continuous variable and capture probability is a smooth function of x (decreasing with distance). To stratify distances into a number of distance intervals and treat capture probability within each interval as constant, is at best an approximation that neglects heterogeneity with the distance intervals. A neat alternative is to treat x as continuous in the model, but in this case animal-level stratification breaks down because there are an infinite number of possible xs. It therefore makes sense to model the process generating x, by way of the pdf $\pi(x)$. For example, in line transect sampling, $\pi(x) = \frac{1}{w}$. We could also use a state model if x was sex or any other discrete variable, but it might be difficult to specify a suitable form for $\pi(x)$. The animal-level stratification option avoids the need to specify $\pi(x)$.

Unobserved heterogeneity When the variables causing heterogeneity are not or cannot be observed, the animal-level stratification approach is not an option: if we don't observe x, we can't estimate $N(x)$. In this case, a state model of some sort is essential for including heterogeneity into inference.

We deal with each of these three approaches below. We consider only the case in which the survey-level variables are constant within each capture occasion. For the most part, we treat the variables causing heterogeneity ($\underline{x}$) as constant across capture occasions, although we do consider some models in which they vary stochastically between capture occasions but are constant within occasion. This sort of model might be appropriate if $\underline{x}$ involved location, and animals could move between surveys.

11.3.1 Notation

Technical section
↓

Here are reminders of some of the key mark-recapture notation that you need to be familiar with to follow the methods of this section:

ω_{si} is the capture indicator for the ith animal on capture occasion s: $\omega_{si} = 1$ if animal i is captured on occasion s, otherwise it is zero.

$\underline{\omega}_{si}$ is the capture history of the ith animal at the start of capture occasion s: $\underline{\omega}_{si} = (\omega_{1i}, \ldots, \omega_{(s-1)i})$.

c_{si} indicates whether animal i has been captured at least once by the start of occasion s: c_{si} is 1 if animal i has been captured, and zero otherwise.

$\underline{0}_s = (0, \ldots, 0)$ (a vector of $(s-1)$ zeros) is used to denote the capture history of an animal which has not been captured by the start of occasion s.

We also define a little new notation:

Ω is the array of the N capture histories in the population; its ith row is the capture history for animal i.

X is the array of the N animal-level variables; its ith row, $\underline{x}_i$, is the vector of animal-level variables for animal i.

11.3.2 Animal-level stratification

Animal-level stratification has been proposed by Chapman and Junge (1956), El Khorazaty *et al.* (1977) and Seber (1982), among others. The full likelihood for S capture occasions when there is no heterogeneity and capture histories are observed is multinomial (see Exercise 6.5). With stratification, the likelihood can be written as a product of multinomial likelihoods (one for each of the K animal-level strata) as follows:

$$
L(N, \underline{\theta}) = \prod_{x=1}^{K} \left(\frac{N(x)!}{(N(x) - n(x))! \prod_{\underline{\omega}_s} n(x, \underline{\omega}_{S+1})!} \right)
$$
$$
\times \prod_{\underline{\omega}_{S+1}} P(x, \underline{\omega}_{S+1})^{n(x, \underline{\omega}_{S+1})} P(x, \underline{0}_{S+1})^{N(x) - n(x)}
$$

$$(11.13)$$

where

$x = 1, \ldots, K$ indexes stratum,

$N(x)$ is the number of animals in stratum x,

$n(x)$ is the number of different animals in stratum x that are captured over the S capture occasions,

$\prod_{\underline{\omega}_{S+1}}$ is the product over all possible observed capture histories,

$n(x, \underline{\omega}_{S+1})$ is the number of captured animals in stratum x with capture history $\underline{\omega}_{S+1}$,

$P(x, \underline{\omega}_S)$ is the probability of observing capture history $\underline{\omega}_{S+1}$ for an animal in stratum x; it has an unknown parameter vector $\underline{\theta}$ that we do not show explicitly.

If the parameter vector $\underline{\theta}$ has no components in common across strata, the survey is equivalent to K separate surveys – one for each stratum. Often, components are common to more than one stratum, for example, if $\underline{\theta}$ contains a parameter relating to survey effort, since the same effort is applied to all animals. Pollock *et al.* (1984), Huggins (1989) and Alho (1990) all model capture probability as a logistic function, depending on one or more of animal-level variables, survey-level variables, and whether or not the animal has previously been captured. We discuss this form of the capture probability function below.

Cormack (1989), Feinberg (1972) and Evans *et al.* (1994), by contrast, develop log-linear capture probability models for mark-recapture surveys with animal–level stratification. Nonlinear constraints on the capture function parameters, $\underline{\theta}$, are used by Evans *et al.* (1994) to give a flexible form that allows capture probabilities to depend on one or more of animal-level variables, survey-level variables, and whether or not the animal has previously been captured.

There are two scenarios in which animal-level stratification is not feasible or at least not ideal. These are when the animal-level variables are not observed, and when they are observed but can take on too many possible values to make stratification feasible. We deal with each in turn below.

11.3.3   Unobserved heterogeneity

Even when we can't observe the variables causing heterogeneity, it is possible to model the heterogeneity. Think of the source of heterogeneity (whatever it is) as dividing the population into strata, as above, with no heterogeneity in them. We can formulate a state model that describes how animals are allocated to the strata probabilistically. For example, if sex was the only source of heterogeneity, a simple state model would specify the probability $\pi(0) = \phi$ that an animal is female ($x = 0$), and hence $\pi(1) = 1 - \phi$ that it is male ($x = 1$). If animals' sexes are independent, then the probability of the sexes of animals $1, \ldots, N$ being $\boldsymbol{X} = (x_1, \ldots, x_N)$ is just

$$\pi(\boldsymbol{X}) \quad = \quad \prod_{i=1}^{N} \pi(x_i) \qquad (11.14)$$

Suppose also for the moment that we have an expression for the probability of observing the matrix of capture histories, $\boldsymbol{\Omega}$, for the case in which we know the sex of every animal in the population. This is our observation model, which we call $P(\boldsymbol{\Omega}|\boldsymbol{X})$, and it has some unknown parameters $\underline{\theta}$. Considered as a function of its parameters, it is a likelihood function. But it is no use on its own because we don't know $\boldsymbol{X}$ and so can't evaluate it. This is where the state model can help. We can use it to average out the unobserved part of the likelihood, leaving the marginal likelihood given what we did observe (namely the capture histories, $\boldsymbol{\Omega}$):

$$L(N, \phi, \underline{\theta}|\boldsymbol{\Omega}) \quad = \quad \sum_{\boldsymbol{X}} \pi(\boldsymbol{X}) P(\boldsymbol{\Omega}|\boldsymbol{X}) \qquad (11.15)$$

where the sum is over all possible combinations of sexes, $\boldsymbol{X}$ that could have given the capture histories we observed. Conceptually, we can evaluate this for every possible set of values for $(N, \phi, \underline{\theta})$ and find the set of parameter values that maximizes the likelihood. This is our maximum likelihood estimate. We can, in principle at least, use a state model to incorporate heterogeneity into our likelihood even when the heterogeneity is unobservable.

Another point to note is that the variables causing heterogeneity do not enter this likelihood explicitly – all we have is a state model for "stratum" or "animal type", and since we don't observe it, this could be any variable, or a combination of lots of variables, and the likelihood would still be valid.

This formulation gives us a very general way of incorporating unobserved heterogeneity of any sort into mark-recapture likelihoods. But if we don't even know what variables are causing heterogeneity, what sort of state model is appropriate? There are two main approaches:

Continuous p

With this approach, first suggested by Burnham (1972), the state model is a probability density function for the capture probabilities themselves. The population is conceived of as containing a potentially infinite number of capture probabilities rather than some finite number of sub-populations with a single capture probability per sub-population. Capture probabilities might arise from a host of animal-level variables, but the argument goes, as these are not observed, it makes sense to model the distribution of capture probabilities $(\pi(p))$ directly. A beta distribution is often used, as it spans the interval 0 to 1, as do probabilities, is quite flexible and has only two parameters. The capture probabilities of the population, $p_1, \ldots, p_N$ are modelled as N independent realizations of $\pi(p)$.

Discrete p

With this approach, developed by Agresti (1994), Norris and Pollock (1995, 1996) and Pledger (2000), the population is conceptualized as containing some finite number (K) of sub-populations with a single capture probability, θ_x in sub-population x ($x = 1, \ldots, K$), and the probability that an animal is in sub-population x is ϕ_x. Capture probabilities $p_1, \ldots, p_N$ are modelled as N independent realizations of a multinomial state model $\pi(x)$, such that the capture probability of animal i (p_i) is equal to θ_x with probability ϕ_x. This model has K parameters; usually K is specified in advance.

The distribution of the observed capture history for the ith animal, $\underline{\omega}_i$ can be thought of as a mixture of K observation models:

$$P(\underline{\omega}_i) = \sum_{x=1}^{K} \phi_x P(\underline{\omega}_i|x)$$

$$= \sum_{x=1}^{K} \phi_x \prod_{s=1}^{S} \theta_x^{\omega_{si}} [1 - \theta_x]^{1-\omega_{si}} \qquad (11.16)$$

where $P(\underline{\omega}_i|x)$ is the observation model for animal i.

As written above, the capture probability function depends only on x, the unobservable sub-population to which animal i belongs, not on survey-level variables or capture history: $p(x) = \theta_x$. Pledger (2000) developed very general models for unobserved heterogeneity by modelling capture probability as a logistic function of trap-response (a function of c_{si}) and capture occasion (s), as well as the unobservable x, by parametrizing $p(x, c_{si}, l_s)$ as follows:

$$\frac{e^{\theta_0 + \theta_{x_i} + \beta_{c_{si}} + \tau_s + (\tau\beta)_{sc_{si}} + (\tau\theta)_{sx_i} + (\beta\theta)_{c_{si}x_i} + (\tau\beta\theta)_{sc_{si}x_i}}}{1 + e^{\theta_0 + \theta_{x_i} + \beta_{c_{si}} + \tau_s + (\tau\beta)_{sc_{si}} + (\tau\theta)_{sx_i} + (\beta\theta)_{c_{si}x_i} + (\tau\beta\theta)_{sc_{si}x_i}}} \qquad (11.17)$$

where the τ_s is the parameter for survey occasion s, $\beta_{c_s i}$ is the trap-response parameter, θ_{x_i} is a random animal parameter, and the parameters in brackets are two-way and three-way interaction terms.

With more general detection function models, like that of Equation (11.17) we get the likelihood:

$$L(N, \underline{\theta}, \underline{\phi}) = \frac{N!}{(N-n)! \prod_{\underline{\omega}_{S+1}} n(\underline{\omega}_{S+1})!} \tag{11.18}$$

$$\times \prod_{i=1}^{N} \sum_{x=1}^{K} \phi_x \prod_{s=1}^{S} p_{si}(x)^{\omega_{si}} [1 - p_{si}(x)]^{1-\omega_{si}}$$

where $p_{si}(x) = p(x, c_{si}, l_s)$, $\prod_{\underline{\omega}_{S+1}}$ is the product over all capture histories that could be observed and $n(\underline{\omega}_{S+1})$ is the number of animals with capture history equal to $\underline{\omega}_{S+1}$

Simpler models are constructed by omitting components of the parameter vector. For example, omitting the parameters in brackets and $\beta_{c_{si}}$ gives a model that allows capture probability to depend on x and survey occasion s only.

↑
Technica
section

11.3.4 Classification of models

Discussion of mark-recapture models M_h, M_{th}, M_{bh} and M_{tbh} was deferred from Chapter 6. Having dealt with capture functions that include animal-level variables, likelihoods for animal-level stratification models, and a likelihood with a state model for animal-level variables, we are now in a position to discuss these models.

As noted in Chapter 6, the classification scheme of Pollock (1974) and Otis *et al.* (1978) for mark-recapture models is a scheme based only on the kind of variables that affect capture probability. In Equation (11.17) we have a model in the class M_{tbh} – one depending on capture occasion (subscript t), trap-response (subscript b) and animal-level variables (subscript h).

By omitting the appropriate parameters from the very general Equation (11.17), we get models M_h, M_{th} and M_{bh}.

Notice that if we omit the interaction terms from Equation (11.17), we still have a model in the class M_{tbh}. (Pledger (2000) uses the notation $M_{t \times b \times h}$ to denote a model with all interaction terms, M_{tbh} to denote a model with no interaction terms in, and so on.)

Notice also that in Equation (11.17), all variables enter as factors (i.e. the explanatory variables are all discrete and there is a parameter for each level of the variable). In some contexts it makes sense for at least some of the variables to be continuous. For example, if on capture occasion s we measured the search/capture effort l_s, we could replace the parameters $\tau_1, \ldots, \tau_S$ with a single parameter τ and have the term $\tau \times l_s$ instead of τ_s in the model. Similarly, when we observe the animal-level variable x (or more generally $\underline{x}$), and it is continuous (distance from the transect line in line transect surveys for example), modelling it as a factor is not appropriate. We deal with this sort of model in some detail below. The classification of Pollock (1974) and Otis $et\ al.$ (1978) is useful, but does not define the model fully – it indicates only classes of model and within each class there are potentially many different models.

11.3.5 Other methods

Other methods which do not require a likelihood have been developed for handling the heterogeneity that most biological populations exhibit. For example for many years, the standard method for fitting model M_h used the jack-knife. The number of distinct animals caught is a very simple but biased estimate of population size, and the jack-knife is used to reduce that bias (Otis $et\ al.$, 1978; Burnham and Overton, 1978, 1979). Pollock and Otto (1983) also developed a jack-knife-type estimator for model M_{bh}. The method of fitting for this model proposed by Otis $et\ al.$ (1978) was a generalized removal method, based on the idea that the mean probability of capture of previously uncaught animals will decrease over time, as the most easily caught animals will tend to be caught first. It is assumed that this probability will stabilize after the first few samples, and this stable probability is estimated by weighted regression, after deleting early samples for which the mean probability is found to be significantly higher.

Chao (1987, 1989) also obtained an estimate of population size by adding a term to the number of distinct animals caught. In her case, this term is the square of the number of animals caught exactly once, divided by twice the number caught exactly twice. The jack-knife estimator performs best when numbers of recaptures are large, and Chao's estimator appears to perform better for small numbers of recaptures.

Other approaches have also been proposed. Perhaps most notable are the coverage estimators. Sample coverage for a given sample is defined as the proportion of the total individual capture probabilities corresponding to that sample accounted for by the captured animals. Chao $et\ al.$ (1992) used this concept to develop an estimator under model M_{th}, and the approach was extended for use with any of the eight models of Otis $et\ al.$ (1978) by Lee and Chao (1994). Ashbridge and Goudie (2000) review estimators for model M_h, introduce new coverage estimators, and assess the performance of various estimators under model M_h by simulation.

Martingale methods, based on the use of estimating equations (Godambe, 1985; Lloyd, 1987), have also been developed. Schwarz and Seber (1999) provide a useful review of the methods.

11.3.6 Observable heterogeneity: Horvitz–Thompson and conditional likelihood

A Horvitz–Thompson estimator of abundance was introduced and used in Chapter 4 (Section 4.3). We returned to it in Chapter 7 (Section 7.2.2), where it was adapted for the case in which inclusion probability is estimated, not known. In this case it has the form

$$\hat{N} = \sum_{i=1}^{n} \frac{1}{\hat{p}_i} \tag{11.19}$$

where $\hat{p}_i$ is the estimated probability that sampling unit i is included in the sample.

While the Horvitz–Thompson estimator proper is unbiased, the Horvitz–Thompson-like estimator of Equation (11.19) is not necessarily so (although it is sometimes asymptotically unbiased). It does, however, provide a way of estimating abundance using mark-recapture methods in the presence of heterogeneity, without having to model the heterogeneity.

Heterogeneity is problematic because we don't know the capture probabilities of animals we did not capture. A very convenient feature of the estimator of Equation (11.19) is that it does not involve inclusion probabilities for animals that were not captured. In order to use Equation (11.19), we need only estimate the capture probabilities of animals in the sample. We saw in Chapter 6 that inference from mark-recapture surveys involves estimating the capture probabilities of animals in the sample. We now describe more general methods than those in Chapter 6 for estimating capture probabilities of animals in the sample in the presence of heterogeneity. They were developed independently by Huggins (1989) and Alho (1990).

The methods use logistic capture functions, similar to Equation (11.17), but with continuous variables rather than factors where appropriate, and possibly without some of the terms in Equation (11.17) (which is the most general form for the case where all explanatory variables are factors).

The method involves estimation of the parameter vector $\underline{\theta}$ and hence capture probability, by maximizing a conditional likelihood, given the animal-level variables and capture histories of captured animals. This MLE can be used with the Horvitz–Thompson-like estimator of Equation (11.19).

Techni◖
section
↓

We write the probability that animal i is captured on at least one of the S capture occasions as

$$p_i = 1 - \prod_{s=1}^{S}[1 - p_{si}] \qquad (11.20)$$

where p_{si} is the probability of catching animal i on occasion s. In general, $p_{si} = p(\underline{x}_i, \underline{\omega}_{si}, \underline{l}_s)$, with an unknown parameter vector $\underline{\theta}$, but we write it as p_{si} here for brevity.

By Bayes' theorem, the conditional probability of observing capture history $\underline{\omega}_{(S+1)i}$, given that animal i was captured at least once (i.e. that $c_{(S+1)i} = 1$) is

$$P(\underline{\omega}_{(S+1)i} \,|\, c_{(S+1)i} = 1) = \frac{\prod_{s=1}^{S} p_{si}^{\omega_{si}} [1 - p_{si}]^{1-\omega_{si}}}{p_i}$$
$$(11.21)$$

and the conditional likelihood for the n observed capture histories is

$$L_{\underline{\omega}|c=1} = \prod_{i=1}^{n} P(\underline{\omega}_{(S+1)i} \,|\, c_{(S+1)i} = 1) \qquad (11.22)$$

Maximizing this likelihood with respect to the parameters gives the MLE $\hat{\underline{\theta}}$, and the MLE of the capture function for animal i is p_{si}, which we refer to as $\hat{p}_{si}$. The probability that animal i is included in the sample is estimated as

$$\hat{p}_i = 1 - \prod_{s=1}^{S}[1 - \hat{p}_{si}] \qquad (11.23)$$

The Horvitz–Thompson-like estimator of abundance is obtained by substituting this in Equation (11.19). Huggins (1989) and Alho (1990) showed that this estimator is asymptotically unbiased and normal, and derived expressions for its asymptotic variance.

↑
Technical
section

11.3.7 Observable heterogeneity: full likelihood models

The Horvitz–Thompson-like estimator described above only works when the population is static with respect to the heterogeneity variable(s). In this case, as long as an animal was captured at least once, its x is known for every capture occasion, and in particular its x is known for occasions when it was not captured. The approach breaks down if x can change between occasions: on occasions we missed the animal, did we miss it by chance, or because it took on a particularly uncatchable x on the occasion? What x should we associate with it in Equations (11.21) and (11.22)?

Similarly, animal-level stratification is only applicable if animal-level variables are static over the survey period.

A full likelihood for a mark-recapture experiment incorporates a state model for x, and with an appropriate model, it can deal with dynamic x as well as static x. It is more general than the Horvitz–Thompson conditional likelihood approach, but the generality has a cost: stronger assumptions about the xs in the population must be made, hence likelihood estimators are likely to be less robust than conditional likelihood estimators. In addition, it may be quite difficult to suggest plausible forms for the state model because in some cases very little will be known about it. Nevertheless, the full likelihood approach is appealing because of its generality and because it extends quite naturally to the sort of approach being used increasingly for inference with open populations.

Conceptually the way we get a likelihood with observed heterogeneity variables is almost identical to that for the case where heterogeneity variables are not observed. In both cases, we use a state model to average out the unobserved heterogeneity variables, but now not all heterogeneity variables are unobserved.

Technical section
↓

Consider the simple example we used before, in which sex is the only variable causing heterogeneity. For convenience, we number animals in the order they are captured, so that the first n animals are captured and are therefore of known sex by the end of the survey. X can then be written as $(X_o, X_u)^T$, where X_o is the sexes of the n captured animals (subscript o for "observed") and X_u is the sexes of the $N - n$ uncaptured animals (subscript u for "unobserved").

Now we average over the unobserved sexes, X_u in much the same way as we did in Equation (11.15) to get the marginal likelihood given what we did observe (namely the capture histories Ω and the sexes X_o):

$$L(N, \phi, \underline{\theta} | \Omega, X_o) = \sum_{X_u} \pi(X_o, X_u) P(\Omega | X_o, X_u)$$

(11.24)

where the sum is over all possible combinations of sexes, X that could have given the capture histories, Ω, and the sexes, X_o, that we observed. Conceptually, we can evaluate this for every possible set of values for $(N, \phi, \underline{\theta})$ and find the set of parameter values that maximizes the likelihood. This is our maximum likelihood estimate.

↑
Technical
section

Recall from Chapter 9 that animal-level variables may be static or may change between survey occasions, and if they change, there are several ways in which they can do so. In the following subsections, we derive mark-recapture likelihoods using state models for static animal-level variables and state models for animal-level variables with Markov dynamics. There are many other possibilities that we do not have space to consider.

We derive the likelihoods using a sequential formulation rather than the global approach of the sort illustrated in Equation (11.24). That is, we average over the unobserved variables at every capture occasion and update the state model for the unobserved variables on each occasion.

Static population; individually identifiable animals

It sometimes helps to think of marking as removing animals from the population of unmarked animals. If animal-level variables (the $\underline{x}$s) don't change over the surveys, then there is no uncertainty about the $\underline{x}$s of marked animals (because they've been seen). The only uncertainty about the $\underline{x}$s is in the unmarked population. So we need a state model for $\underline{x}$ for the unmarked population only.

Recall from Sections 5.2.4 and 6.3.5 that heterogeneity tends to result in the more detectable animals being marked (or removed) first. So the distribution of $\underline{x}$s of unmarked animals changes after every marking (or removal) occasion; the average catchability of uncaptured animals drops. An appropriate state model must reflect this. That is, $\pi(\underline{x})$ must depend on both occasion s and the capture function $p(\underline{x}, \underline{\omega}, l)$. We derive a state model of this sort in Appendix D. The state model is the pdf of $\underline{x}$, given that an animal is uncaptured by occasion s (i.e. that it has capture history $\underline{0}_s$). We refer to it as $\pi(\underline{x}|\underline{0}_s)$; details of its derivation are left to the appendix.

We use this state model to average out the uncertainty in the capture probabilities of (as yet) unmarked animals and so get a likelihood that can be evaluated given the observed data.

Technical
section
↓

Using our state model, we can evaluate the expected probability of capturing (or missing) an animal with unknown $\underline{x}$ on occasion s. It turns out to be more convenient to work with the expected probability of missing an animal than of capturing an animal. We refer to this as $E\left[P(\underline{0}_{(s+1)}|\underline{x})\right]$, where $P(\underline{0}_{(s+1)}|\underline{x})$ is the probability that an animal is uncaptured by the start of capture occasion $s+1$ (given $\underline{x}$ and the survey-level variables applied on each survey). In much the same way as we obtained the likelihood Equations (7.10) and (11.2), we can take expectation over the unknown $\underline{x}$s in the unmarked population on occasion s to get the likelihood for unmarked animals.

It is convenient to number animals in the order they are first captured. At the start of capture occasion s, animals $1, \ldots, M_s$ are marked and animals $(M_s + 1), \ldots, N$ are unmarked. The likelihood component for the $U_s = N - M_s$ unmarked animals, of which $u_s = n_s - m_s$ are captured on survey s, is

$$
L_{s0} = \binom{U_s}{u_s} E[P(\underline{0}_{(s+1)} | \underline{0}_s, \underline{x})]^{U_s - u_s}
$$
$$
\times \prod_i p(\underline{x}_i, \underline{0}_s, \underline{l}_s) \pi(\underline{x}_i | \underline{0}_s)
$$

$$(11.25)$$

where the product is over the indices of all animals with capture history $\underline{\omega}_s$ that are captured on occasion s.

If animals are marked in a way that makes them individually identifiable, the component of the likelihood for the marked animals on capture occasion s only is

$$
L_{s1} = \prod_{i=1}^{M_s} p(\underline{x}_i, \underline{\omega}_{si}, \underline{l}_s)^{\omega_{si}} [1 - p(\underline{x}_i, \underline{\omega}_{si}, \underline{l}_s)]^{1-\omega_{si}}
$$

$$(11.26)$$

The likelihood for capture occasion s, given that animals $i = 1, \ldots, M_s$, with associated animal-level variables $\underline{x}_1, \ldots, \underline{x}_{M_s}$ are marked, is just

$$
L_s = L_{s0} \times L_{s1} \tag{11.27}
$$

The full likelihood is the product of these L_ss for $s = 1, \ldots, S$. This simplifies to

$$
L = \binom{N}{n} E[P(\underline{0}_{(S+1)} | \underline{x})]^{N-n} \times \prod_{i=1}^{n} \pi(\underline{x}_i) \times \tag{11.28}
$$
$$
\prod_{i=1}^{n} \prod_{s=1}^{S} p(\underline{x}_i, \underline{\omega}_{si}, \underline{l}_s)^{\omega_{si}} [1 - p(\underline{x}_i, \underline{\omega}_{si}, \underline{l}_s)]^{1-\omega_{si}}
$$

where n is the number of different animals marked over all S capture occasions (recall that u_s is the number of unmarked animals captured on occasion s).

↑
Technical
section

If animals are not individually identifiable but are identifiable only by their $\underline{x}$s, which can be one of K possible values $\underline{x}_1, \ldots, \underline{x}_K$, then the likelihood is slightly different (see Exercise 11.3).

Independent dynamics

If detectability depends on any animal-level variable that can change between capture occasions, a state model with static $\underline{x}$ will not be appropriate. If $\underline{x}_i$ is determined on each occasion independently of its values on other occasions, what we call a state model with independent dynamics is appropriate. In the woodland/grassland example, this would be the case if the probability that an animal is in woodland on capture occasion s is ϕ on every occasion, no matter where it was on previous occasions. Because $\underline{x}_{si}$ is independent of all previous $\underline{x}_{s^*i}$ ($s^* < s$), the conditional probability density of $\underline{x}_{si}$, given capture history $\underline{\omega}_{si}$ is just $\pi(\underline{x}_{si})$. That is,

$$\pi(\underline{x}_s | \underline{\omega}_s) \ = \ \pi(\underline{x}) \tag{11.29}$$

The likelihood equations developed for the static $\underline{x}$ scenario above apply, but with $\pi(\underline{x}_s | \underline{0}_s)$ replaced by $\pi(\underline{x})$, a pdf that remains unchanged over all survey occasions.

Population level drift

If $\underline{x}$ is independent between capture occasions, but the state model changes between occasions, we have what we call a population level drift model. In the woodland/grassland example, this would be the case if the probability that an animal is in woodland on capture occasion s is independent of where it was on previous occasions, but the probability that it is in woodland changes over time. In this case, the likelihood equations developed for the static $\underline{x}$ scenario above apply, but with $\pi(\underline{x}_s | \underline{0}_s)$ replaced by

$$\pi(\underline{x}_s | \underline{\omega}_s) \ = \ \pi(\underline{x}_s) \tag{11.30}$$

in which the pdf of $\underline{x}_s$ depends on s through parameters of the state model, not marking or removal.

Markov dynamics

Finally, consider the case where the $\underline{x}$ associated with an animal on occasion s depends on $\underline{x}_{(s-1)i}$, but given $\underline{x}_{(s-1)i}$, not on any previous $\underline{x}_{s^*i}$; $s^* < (s-1)$.

In this case, the conditional pdf of animal is $\underline{x}$ on occasion s depends not only on its capture history, but on its history of $\underline{x}$s. In Appendix D we derive $\pi(\underline{x}_{si}|\mathcal{H}_{si}^o)$, the conditional pdf of $\underline{x}_{si}$ on occasion s, given the animal's observed history of $\underline{x}$s prior to occasion s, which we refer to as $\mathcal{H}_{si}^o$.

The likelihood for unmarked animals on occasion s in this case is the same as that for the static population case developed above, but with $\pi(\underline{x}_{si}|\mathcal{H}_{si}^o)$ in place of $\pi(\underline{x}_s|\underline{\omega}_s)$.

The likelihood for animals that have been captured at least once by the start of occasion s is different from the static population case because the $\underline{x}$s on occasion s of previously captured animals are random variables and are unknown unless they are observed on occasion s. (They could have changed since being observed previously.) To evaluate the likelihood, we average over the unobserved $\underline{x}$s as follows.

Technical section
↓

Likelihood Equation (11.26) is an observation model, conditional on the $\underline{x}_{si}$s of previously captured animals; call it $L_{s1}(\underline{\theta})$. Considered as a likelihood function, our state model for these $\underline{x}_{si}$s is

$$L_{s1}(N, \underline{\phi}) \;=\; \prod_{i=1}^{N} \pi(\underline{x}_{si}|\mathcal{H}_{si}^o) \qquad (11.31)$$

To evaluate the likelihood of the observed data for previously captured animals, we average out the unobserved $\underline{x}_s$s in $L_{s1}(\underline{\theta}) \times L_{s1}(N, \underline{\phi})$ to get:

$$L_{s1} \;=\; \prod_{i=1}^{M_s} E[P(0|\mathcal{H}_{si}^o)]^{1-\omega_{si}} p(\underline{x}_{si}, \underline{\omega}_{si}, \underline{l}_s)^{\omega_{si}} \pi(\underline{x}_{si}|\mathcal{H}_{si}^o)$$

$$(11.32)$$

where $E[P(0|\mathcal{H}_{si}^o)]$ is the mean probability of missing an animal with history $\mathcal{H}_{si}^o$ (see Appendix D for details).

↑
Technical section

11.4   Removal methods

11.4.1   Animal-level stratification

Otis *et al.* (1978) developed what they called "general removal" models for animal-level stratification. Evans *et al.* (1994) developed sophisticated log-linear models as alternatives to the models of Otis *et al.* (1978). Schwarz and Seber (1999) review recent developments in this area.

11.4.2   Horvitz–Thompson and conditional likelihood

Horvitz–Thompson-like estimators can be used with removal methods in much the same way as for mark-recapture surveys (Huggins and Yip, 1997). As there, they have the advantage over full likelihood methods of not requiring assumptions about the xs of the unobserved animals, and the disadvantage that they are only applicable with a static state model for x.

The method involves estimation of the parameter vector $\underline{\theta}$, and hence capture probability, by maximizing a conditional likelihood, given the animal-level variables and capture histories of removed animals.

Technical section ↓

Because animals are removed when they are captured, the probability that animal i is captured some time in the S capture occasions is

$$p_i = 1 - \prod_{s=1}^{S}[1 - p_{si}] \qquad (11.33)$$

where p_{si} is defined as in Section 11.3.6.

The conditional probability of observing capture history $\underline{\omega}_{(S+1)i}$ for animal i, given it was captured at all (i.e. that $c_{(S+1)i} = 1$), is

$$P(\underline{\omega}_{(S+1)i}|c_{(S+1)i} = 1) = \frac{\prod_{s=1}^{S}\left\{p_{si}\prod_{s^*=1}^{s-1}(1 - p_{s^*i})\right\}^{\omega_{si}}}{p_i}$$

$$(11.34)$$

and the conditional likelihood for the n observed capture histories is

$$L_{\underline{\omega}|c=1} = \prod_{i=1}^{n}P(\underline{\omega}_{(S+1)i}\,|\,c_{(S+1)i} = 1) \qquad (11.35)$$

Maximizing this likelihood with respect to the parameters gives the MLE of the parameters $\hat{\underline{\theta}}$, and hence the MLEs, $\hat{p}_{si}$, of the capture probabilities p_{si} ($s = 1, \ldots, S;\ i = 1, \ldots, N$). Substituting these in Equation (11.19), we get the Horvitz–Thompson-like estimator of abundance. Huggins and Yip (1997) showed that this estimator is asymptotically unbiased and normal, and derived expressions for its asymptotic variance.

↑ Technical section

11.4.3 Full likelihood methods

Recall that mark-recapture surveys contain within them removal surveys – each marking occasion removes some animals from the population of unmarked animals. Generalizations of the simple removal and catch-effort methods of Chapter 5 to include animal-level heterogeneity are therefore contained in the development of mark-recapture models with heterogeneity above. Specifically, the relevant likelihood functions for each survey occasion are the L_{s0}s above. Because captured animals are removed, the L_{s1}s do not apply. A general removal method likelihood for heterogeneous populations is therefore

$$ L \;=\; \prod_{s=1}^{S} L_{s0} \tag{11.36} $$

The change-in-ratio methods dealt with in Chapter 5 treat the numbers of animals of each type in the initial population (the $N_1(\underline{x})$s) as parameters to be estimated, while the methods of incorporating heterogeneity covered in this chapter treat these numbers as random variables: realizations of some underlying state whose parameters are to be estimated. The change-in-ratio methods of Chapter 5 are applicable only if the population is static and the $N_1(\underline{x})$s are treated as parameters. In the vast majority of applications of change-in-ratio methods in the literature, this is an appropriate approach. A common example is when $\underline{x}$ is a scalar indicating sex.

The developments of this chapter generalize change-in-ratio methods in two respects that may be useful in practice. The first is in allowing the $N_1(\underline{x})$s to be random variables. If the heterogeneity variable $\underline{x}$ is such that it can take on many values, a model which treats the $N_1(\underline{x})$s as random realizations of an underlying state model may involve substantially fewer parameters. This might give improved precision in estimating N. The second potentially useful extension is in allowing dynamic heterogeneity. The independent dynamics case does not seem appropriate as a change-in-ratio method because it undermines the basis of the method – the numbers of each type present on each occasion are completely independent of the numbers of each type that have been removed. A model with Markov dynamics might, however, be useful. This might be the case if $\underline{x}$ indicated region or stratum and there was some movement between regions. It might also be useful if $\underline{x}$ was size class and the surveys spanned sufficient time that animals moved up through the size classes.

11.5   Summary

We can arrange the models of this chapter into a hierarchy based on state models. At the bottom are those that do not involve state models (mark-recapture models based on animal-level strata with observable animal-level variables); then there are those that involve state models to integrate out heterogeneity associated with unobserved animals (distance sampling models with covariates; mark-recapture and removal models with observed heterogeneity); finally there are those in which all heterogeneity associated with animals is integrated out using a state model (mark-recapture models with unobserved heterogeneity).

In some closed-population contexts, a full likelihood approach involving a state model might be preferable, in others not. State models are usually an integral part of open-population models (explicitly or otherwise).

The following are some advantages of using state models in a closed-population context.

- Inference includes inference about the state model, and this may be of interest in itself. The spatial distribution of animals in the survey region is one example; the distribution of group sizes is another.

- Unlike the Horvitz–Thompson method, estimation based on state models can accommodate dynamic processes.

- The full likelihood approach, with state model, provides a unifying conceptual framework that naturally links various apparently different methods. In Chapter 12 we explore some links between closed population models; in Chapter 13 we demonstrate how the formulation with state models links naturally to formulations for open population models.

Disadvantages of using state models include the following.

- If the state model is inappropriate, inferences may be misleading. The less is known about what form of state model is appropriate, the more this is a problem. Horvitz–Thompson-like estimators or other estimators that are not based on a state model might be more robust.

- The number of parameters to be estimated can be large if the state model is multi-dimensional.

Note that the generalizations of this chapter incorporate as special cases, the simple removal, catch-effort and change-in-ratio methods that were presented in Chapter 5.

11.6 Exercises

Exercise 11.1 Derive the distance sampling likelihood with covariates, given in Equation (11.2) by

(i) specifying a state model, $\pi(\underline{x})$, where $\underline{x} = x, z$ and z is group size (assume independence between x and z, and do not assume any specific form for the pdf of z),

(ii) specifying the observation model (a product N of binary pdfs), and

(iii) integrating out the unobserved animal-level variables.

Exercise 11.2 Show that in the case of line transect sampling, the Horvitz–Thompson-like estimator of group abundance given in Equation (11.11) can be written as

$$\hat{N}_{grp} = \frac{A}{2L} \sum_{i=1}^{n} \hat{f}(0|z_i, \underline{y}_i, l_i) \qquad (11.37)$$

where L is line length, x is perpendicular distance, z_i is group size, $\underline{y}_i$ is a vector of other animal-level variables, l_i is a survey-level variable, and $\hat{f}(0|z_i, \underline{y}_i, l_i)$ is the estimated intercept of the conditional pdf of x, given z_i, $\underline{y}_i$ and l_i.

Exercise 11.3 Derive a likelihood for a mark-recapture study of a static population with animal-level heterogeneity, in which animals are not individually identifiable but are identifiable only by their $\underline{x}$s, which can be one of K possible values $\underline{x}_1, \ldots, \underline{x}_K$.

Exercise 11.4 Suppose that capture probability in a mark-recapture experiment depends only on sex (not on capture history or survey-level variables). Let $\boldsymbol{X} = (x_1, \ldots, x_N)$ represent the sexes of the N animals in the population, where $x_i = 0$ if animal i is female, and $x_i = 1$ if animal i is male. Let $\boldsymbol{\Omega}$ be the capture history matrix of all N animals. Let $P(\boldsymbol{\Omega}|\boldsymbol{X})$ be the probability of observing $\boldsymbol{\Omega}$, given $\boldsymbol{X} = (x_1, \ldots, x_N)$.

(a) Assuming animals are captured independently of one another, write down an observation model giving the probability of observing the capture history matrix $\boldsymbol{\Omega}$, given the sexes of the N animals in the population.

(b) Assuming animals' sexes are independent and that the probability of being female is ϕ, write down a state model for the sexes of the N animals in the population, $\boldsymbol{X} = (x_1, \ldots, x_N)$.

(c) Using (a) and (b), derive the marginal probability of observing the capture history matrix $\boldsymbol{\Omega}$ if the sex of the n captured animals is not observed.

(d) Using (a) and (b), derive the marginal probability of observing the capture history matrix Ω and the sexes X_o, if the sex of the n captured animals is observed.

Exercise 11.5 Suppose you conduct a two-sample survey of a population of $N = N_1$ animals consisting of $N_1(0)$ females and $N_1(1)$ males at the start of the first survey occasion. Females are detected with probability $p(0, s)$ on survey occasion s, and males with probability $p(1, s)$. Suppose also that the probability animal i is female is $\pi(0) = \phi$ (and the probability that it is male is $\pi(1) = 1 - \phi$) and that animals' sexes are determined independently of each other.

(a) Write down a state model for $N_1(0)$, the number of females in the population, treating $N = N_1 = N_1(0) + N_1(1)$ and ϕ as parameters.

(b) Write down an observation model for $n_1(0)$ and $n_1(1)$, the number of females and males that are detected on the first survey, given $N_1(0)$ and $N_1(1)$.

(c) By multiplying the observation model and state model, and summing over all possible $N_1(0)$ and $N_2(0)$ that could have led to the observed data $n_1(0)$ and $n_1(1)$, show that the probability of detecting $\underline{n}_1 = (n_1(0), n_1(1))'$ is

$$
P(\underline{n}_1) = \left(\frac{N!}{(N - n_1)! \prod_{x=0}^{1} n_1(x)!} \right)
$$
$$
\times (1 - E[p_1])^{N-n_1} \prod_{x=0}^{1} \{\pi(x)p(x, 1)\}^{n_1(x)}
$$

(11.38)

where $n_1 = n_1(0) + n_1(1)$ and $E[p_1] = \sum_{x=0}^{1} p(x, 1)\pi(x)$ is the mean probability (over sex) of detecting an animal on the first survey occasion.

(d) Given that $R_2(0)$ females and $R_2(1)$ males were removed from the population before the start of the second survey occasion, write down a state model for $N_2(0)$, the number of females in the population at the start of this occasion, treating N and ϕ as parameters.

(e) Hence show that

$$
P(\underline{n}_2) = \left(\frac{(N - R_2)!}{(N - R_2 - n_2)! \prod_{x=0}^{1} n_1(x)!} \right)
$$

$$\times (1 - E[p_2])^{N-n_2} \prod_{x=0}^{1} \{\pi(x)p(x,2)\}^{n_2(x)}$$

$$(11.39)$$

where $n_s = n_2(0) + n_2(1)$ and $E[p_2] = \sum_{x=0}^{1} p(x,1)\pi(x)$ is the mean probability (over sex) of detecting an animal on the second survey occasion.

Exercise 11.6   Consider the same scenario as in Exercise (11.5), but suppose that the removed animals are those detected on the first survey occasion (so that $R_2(x) = n_1(x)$ for $x = 0, 1$).

(a) Write down the state model for an animals' sex on the second occasion – i.e. the probability density of x, given that the animal survived the first capture occasion ($x = 0$ for females, $x = 1$ for males).

(b) Show that in this case,

$$P(\underline{n}_2) = \left(\frac{(N - n_1)!}{n_2(0)!n_2(1)![N - n]!} \right)$$

$$\times \prod_{x=0}^{1} \{\pi(x)[1 - p(x,1)]p(x,2)\}^{n_2(x)}$$

$$\times (1 - E[p_2])^{N-n_2} \qquad (11.40)$$

where $n = \sum_{s=1}^{2} \sum_{x=0}^{1} n_s(x)$ is the number of animals captured in all.

(c) Hence show that the likelihood for N, ϕ and $p(x, s)$ is

$$L(N, p(x,s), \phi) = \left(\frac{N!}{(N-n)! \prod_{s=1}^{2} \prod_{x=0}^{1} n_s(x)!} \right)$$

$$\times \prod_{x=0}^{1} \{\pi(x)p(x,1)\}^{n_1(x)}$$

$$\times \prod_{x=0}^{1} \{\pi(x)[1 - p(x,1)]p(x,2)\}^{n_2(x)}$$

$$\times (1 - E[P_2])^{N-n} \qquad (11.41)$$

where $E[P_2] = \sum_{x=0}^{1} \{1 - \prod_{s=1}^{2} [1 - p(x,s)]\}\pi(x)$ is the mean probability (over sex) of detecting an animal on at least one of the two survey occasions.

12
Integrated models

Key idea: animal abundance estimation methods are often not as different as they appear.

Number of surveys:	Single or multiple; depends on method.
State model:	Spatial model for animal distribution in space and/or temporal model for animal distribution in time and/or other state model.
Observation model:	Various; depends on method.

12.1 Introduction

The theme of this chapter is integration. We describe three models that involve integration of some models covered in previous chapters, and very briefly suggest some others. The three integrated models are as follows.

Spatial modelling with distance sampling This integrates the spatial modelling methods for plot surveys of Section 10.3 with distance sampling with covariates (Section 11.2).

Line transect surveys with uncertain detection on the line This integrates distance sampling methods (Chapter 7 and Section 11.2) and a mark-recapture method with a static heterogeneous population (Section 11.3.7).

Migration surveys with uncertain detection This integrates the migration count method (Section 10.2) and a mark-recapture method with a static heterogeneous population (Section 11.3.7).

12.2 Distance sampling with a spatial model

We describe methods developed primarily by Sharon Hedley here; similar methods, but using a different approach, are being developed by Mark Bravington (Bravington, unpublished) at the time of writing this. We do not cover them.

12.2.1 Full likelihoods

Distance sampling likelihoods are naturally formulated in terms of perpendicular distances from transects or radial distances from points, but it usually makes sense to formulate spatial state models in terms of a coordinate system that is independent of the locations of transect lines or points. Full distance sampling likelihoods involve both the spatial state coordinate system and distances from points or lines and as a result they usually involve transformation of distances from lines or points into the spatial coordinate system.

Given the locations of the observer (i.e. the sampled transects or the points), the distance x of an animal from the observer is a known function of animal location (u, v). To illustrate with a simple example, consider a line transect survey with a single line perpendicular to the u-axis at a distance u_l along it. In this case $x = |u - u_l|$. In a point transect survey, the distance of an animal at (u, v) from an observer at (u_l, v_l) is $x = \sqrt{(u - u_l)^2 + (v - v_l)^2}$. In general, the distance of the animal from the observer is some known function, which we write as $x(u, v)$. Note that x denotes distance from transect line or point while $\underline{x}$ denotes the animal-level variables associated with an animal, which include location coordinates (u, v).

Count model

Consider the example we looked at in Section 11.2.1, in which animals are uniformly distributed in the survey region and detection probability depends on distance (x) and group size (z). Writing

Techn
sectio
↓

the detection function in terms of $\underline{x} = (u, v, z)$ as $p(\underline{x})$, Equation (11.2) can be generalized to

$$L(N, \underline{\theta}, \underline{\phi}) = \binom{N}{n} (1 - E[p])^{N-n} \prod_{i=1}^{n} p(\underline{x}_i) \pi(\underline{x}_i)$$

$$(12.1)$$

Here

$$E[p] = \int \int p(\underline{x}_i) \pi(\underline{x}_i) \, du \, dv \qquad (12.2)$$

and integration is over the covered region.

To see that Equation (11.2) is a special case, note that we can write the probability density of $\underline{x} = (u, v, z)$ as

$$\pi(\underline{x}) = \Pr\{(u, v) \text{ is in covered region}\}$$
$$\times \pi(\underline{x} \,|\, (u, v) \text{ is in covered region}) \qquad (12.3)$$

and if animals are uniformly distributed in the covered region, then $\Pr\{(u, v) \text{ is in covered region}\} = \frac{a}{A} = \pi_c$. Suppose we change variables from (u, v) to (x, y) where for line transects, y is distance along the line, and for point transects, y is the angle of the animal from the point (relative to some arbitrarily defined origin). Then $\pi(x, y, z) = \pi_c \pi(x) \pi(y)$ if animals are uniformly distributed in the survey region. Because neither detection probability nor animal density depend on y, $\pi(y)$ can be omitted from the likelihood with no loss, and Equation (12.1) reduces to Equation (11.2).

The maximum likelihood estimate $\hat{\underline{\phi}}$ obtained from the likelihood Equation (12.1) provides an estimate of the spatial distribution of animals, namely $\hat{\pi}(u, v) \equiv \pi(u, v; \hat{\underline{\phi}})$.

Note that Equation (10.6) is also a special case of Equation (12.1), which is obtained when $p(\underline{x}) = 1$ in the covered region (certain detection) and $p(\underline{x}) = 0$ elsewhere.

Nearest neighbour model

↑
Technical
section

By considering the "waiting areas" between detections, Hedley (2000) and Hedley *et al.* (1999) developed an alternative form of likelihood for line transect surveys. It has its roots in waiting time or lifetime models and can be thought of as a kind of nearest neighbour method for line transects with a non-uniform spatial state model. The key idea is that larger areas will be surveyed without detecting any animals in low-density regions than in high-density areas; the area surveyed between consecutive detections contains information about animal density. We summarize the key results below; see Hedley (2000) and Hedley et al. (1999) for more details.

Technical section ↓

Let d be the area surveyed along a transect line from some start point (u_0, v_0) until an animal is detected. Assume that animals are distributed in the survey region according to an inhomogeneous Poisson process with rate parameter $D(u, v)$ at point (u, v). An animal at (u, v), with animal-level variables $\underline{x} = (u, v, z)$, is detected with probability $p(\underline{x}, l)$, where l gives the survey-level variables as a function of location. It follows that the observed process is an inhomogeneous Poisson process with rate parameter $D(u, v)p(\underline{x}, l)$ (see Cressie, 1998, pp 625–626). Hence if d is the distance along the line from (u_0, v_0) to some point (u_d, v_d), then its cumulative distribution function is

$$
\begin{aligned}
F(d \,|\, w, (u_0, v_0)) &= 1 - \exp\left\{-\int\int D(u, v)p(\underline{x}, l)\, du\, dv\right\} \\
&= 1 - \exp\left\{-\lambda((u_0, v_0), d, w)\right\} \quad (12.4)
\end{aligned}
$$

where integration is over all (u, v) in a strip of half-width w centred on the transect line, starting at (u_0, v_0) and ending at (u_d, v_d).

Differentiating gives the density of d. It can be expressed most conveniently using x (distance perpendicular to the line) and y (distance along the line):

$$
\begin{aligned}
f(d \,|\, w, (u_0, v_0)) &= \int_{-w}^{w} D(x, d)p(\underline{x}, l)\, dx \\
&\quad \times \exp\left\{-\lambda((u_0, v_0), d, w)\right\} \quad (12.5)
\end{aligned}
$$

where $D(x, d)$ is $D(u_d, v_d)$ expressed in terms of distance perpendicular to the transect line (x) and distance along the transect line $(y = d)$, and integration is over a line from a perpendicular distance w to the left of the transect line to a perpendicular distance w to the right of the transect line. Note that if $D(u_d, v_d)$ has a constant value $D(d)$ along this line (i.e. it does not vary across the width of the strip), we can write

$$
\begin{aligned}
f(d \,|\, w, (u_0, v_0)) &= D(d)2\mu(d) \\
&\quad \times \exp\left\{-\lambda((u_0, v_0), d, w)\right\} \quad (12.6)
\end{aligned}
$$

where $\mu(d)$ is the effective search half-width at distance d along the transect line. A similar result holds when density changes linearly across the width of the strip (see Exercise 12.3).

Given the waiting distances $d_1, \ldots, d_n$ to the n detections, the likelihood for $\underline{\theta}$, $\underline{\phi}$ and N is

$$L(N, \underline{\theta}, \underline{\phi}) \quad = \quad \prod_{i=1}^{n} f(d_i, w \mid (u_{i-1}, v_{i-1})) \qquad (12.7)$$

For simplicity, we have neglected that part of the likelihood corresponding to the last (truncated) waiting distance, that from the last detection to the end of the searched line (see Hedley *et al.*, 1999; Hedley, 2000, for details).

↑
Technical
section

12.2.2 GLM- and GAM-based methods

The likelihood Equations (12.1) and (12.7) provide a basis for estimating the distribution of animals in the survey region ($\pi(u, v)$) as well as abundance, when animals are not uniformly distributed. However, a practical obstacle to using these methods is that the form of $\pi(u, v)$ will often be both unknown and quite complex, as illustrated in Figure 3.3, for example. Conventional maximization methods require that the form be specified.

Hedley (2000) and Hedley *et al.* (1999) developed more tractable methods for estimating spatial distribution and abundance from distance sampling surveys. While they are not full likelihood methods, they have some advantages:

(1) By using generalized linear models (GLMs) or generalized additive models (GAMs), they accommodate arbitrarily complicated forms for $\pi(u, v)$ without the need to specify the form in advance.

(2) They are readily implemented using standard statistical packages.

The methods parallel the likelihoods of Equations (12.1) (based on counts) and (12.7) (based on waiting distances). The "count method" is perhaps more naturally suited (with minor adaptation) to point transect surveys, although it was not originally developed for this purpose. This is because it involves breaking the covered region into small "segments", something which occurs naturally with point transect surveys (in which the covered region around a point is the "segment"), but not with line transect surveys. The "waiting distance method" is based on the area covered while "waiting" for the next detection. It is perhaps a more natural method for line transect surveys as these involve continuous search along transect lines.

Both methods involve estimation of the detection function parameters, $\underline{\theta}$, from the conditional likelihood, as outlined in Section 11.2.3. This involves an implicit assumption that the distribution of distance x is known within the covered region, and typically that it is uniform. While this is strictly inconsistent with a non-uniform spatial distribution, the range of xs in the covered region in distance surveys is typically such a small fraction of the survey region that the assumption that $\pi(x)$ is uniform locally within the covered region is likely to be very close to truth. If the truncation distance w is such that density is likely to change appreciably within the covered region, this will not be the case.

12.2.3   Confidence intervals

Nonparametric bootstrap methods based on survey sampling units (transects or points) are not ideal because they do not preserve the spatial structure of the sample. In any given resamples, some lines or points appear more than once, and others not at all, so that the spatial spread in each resample tends to be less than that of the original sample. Hedley (2000) and Hedley et al. (1999) developed a parametric bootstrap method and jack-knife method for estimating variance and confidence intervals. The variance estimates are, however, conditional on the estimated effective strip half-width, $\hat{\mu}$. They incorporate this uncertainty using the "delta method" (Seber, 1982), having separately estimated the variance of $\hat{\mu}$. While the methods can work well in some cases, further work is required before a reliable general confidence interval estimation method for these methods is available.

12.2.4   Example: Antarctic minke whales

The International Whaling Commission (IWC) has conducted shipboard line transect surveys of Antarctic minke whales since the 1978/79 austral summer. The first 18 of these fell under the IWC's International Decade of Cetacean Research (IDCR) programme, with each annual survey covering roughly a sixth of the Southern Ocean south of 60°S. The surveys are routinely analysed using conventional line transect methods, of the sort described in Section 7.2. Because minke whale density shows considerable trend with latitude (with highest densities in the south, close to the ice edge) and with longitude, the survey region is divided into strata. Estimates of minke whale abundance are obtained for each stratum. Branch and Butterworth (2001) summarize the methods used and estimates obtained.

Hedley et al. (1999) reanalysed the 1992/93 IWC/IDCR survey, using the spatial modelling methods described above. The location of strata, transects, and detections is shown in Figure 12.1. We summarize results from the count model here. Minke abundance and distribution was estimated

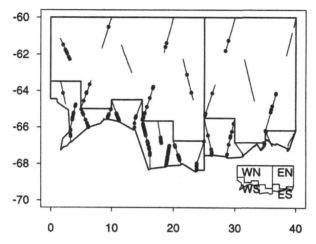

Figure 12.1. The location of transect lines and detected minke whale schools on the 1992/93 IWC/IDCR Antarctic survey. There were four strata, "west-north" (WN), "west-south" (WS), "east-north" (EN) and "east-south" (ES), as indicated in the bottom right of the figure.

using a GAM with log link. The detection function was estimated prior to fitting the GAM. Model selection based on AIC led to latitude, longitude and distance from the ice edge being chosen. The estimated minke school density is shown in Figure 12.2 and the results are compared with the stratified line transect estimates obtained by Borchers and Cameron (unpublished) in Table 12.1.

The point estimates of total abundance from the two methods are remarkably similar and the spatial model provides much more detailed information on the distribution of the animals than does a stratified approach. It also provides information about which are important explanatory variables for distribution. In this example, only latitude, longitude and distance from ice edge were available, but the methods have the potential to related whale distribution to environmental variables, which is useful for future management if global warming changes the Antarctic environment.

The count method estimates of CV are considerably lower than those from a stratified approach. This is what we would expect in general, because the method is able to "explain" changes in density in terms of the observed explanatory variables using relatively few parameters, whereas the stratified approach assigns all changes in density within strata to variance. In this example, however, we need to be cautious not to over-interpret the improved precision of the count method, because there is evidence that the bootstrap-based CV estimate is negatively biased. It is based on an assumption that residuals are independent whereas there is evidence of unmodelled spatial correlation in the results; the dispersion parameter estimated from the original data is 2.68, while the mean estimated dispersion parameter

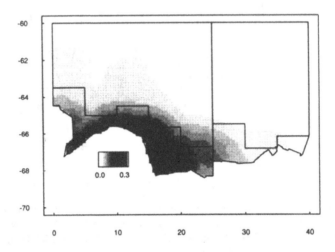

Figure 12.2. The estimated density surface of minke schools from the 1992/93 IWC/IDCR Antarctic survey data.

Table 12.1. Estimated minke school abundance ($\hat{N}$) from the 1992/93 IWC/IDCR Antarctic survey. Estimates are shown from a conventional stratified analysis and from the count method for modelling spatial distribution from line transect data. $\%\widehat{CV}$ is the estimated percentage coefficient of variation (estimated standard error divided by point estimate).

	Stratified		Count model	
Stratum	$\hat{N}$	$\%\widehat{CV}$	$\hat{N}$	$\%\widehat{CV}$
WN	4,810	40.1	4,621	19.9
EN	1,460	49.5	819	28.9
WS	7,412	25.1	8,415	17.6
ES	636	44.3	885	25.1
Total	14,318	23.0	14,740	16.0

across the bootstrap samples is 1.68. While there remains some development to be done before general and reliable confidence interval estimation methods are available, spatial modelling methods can be expected to improve precision by modelling trend within strata. In addition, the methods will improve precision of abundance estimates within sub-regions of the survey region (strata, for example) because they use data from the whole survey region to determine the density in the sub-region, whereas stratified methods use only data from within the sub-region.

12.3   Double-platform distance sampling

Recall that of the methods with uncertain detection probability, distance sampling is the only one that is feasible without more than one survey. The reason for this is that it involves a stronger assumption about detection probability than do other methods, namely that detection probability is known (and equal to 1) at distance zero from the line or point. In some applications, detection of animals at distance zero is not certain. Line transect surveys of marine mammals are a case in point – animals at distance zero may be missed because they are underwater some of the time. The assumption may also be violated on some aerial surveys. When it is violated, a single survey gives an estimate only of relative abundance – it estimates $p(0)N$, where $p(0)$ is the unknown probability of detecting an animal at zero distance from the line or point.

The most successful methods of accommodating uncertain detection at distance zero in distance sampling surveys involve a mixture of distance sampling and mark-recapture methods. Two observers survey the same covered region independently of one another, recording data that allow the analyst to identify animals that are detected by both. Conceptually each observer represents a capture occasion and animals detected by both are recaptures (usually called "duplicates" or "duplicate detections" in this context). Distance from the line or point (x) is a key animal-level variable (because detection probability almost certainly depends on it), but other animal-level and survey-level variables may also be important. Distance from the line or point is special in another respect: $\pi(x)$ can usually be treated as known (as with all distance sampling methods).

In the technical section below, we develop full and conditional likelihood functions and estimators for this sort of model.

12.3.1   Full likelihood

We assume that animals are independently distributed in the survey region. Consider first the case in which observer is the only survey-level variable, and l indexes observer, so that $p(\underline{x}, 1)$ is the

Technical
section
↓

probability that observer 1 detects an animal with animal-level variables $\underline{x}$, and $p(\underline{x}, 2)$ is the probability that observer 2 detects it. If detections are independent, the probability that at least one of them detects it is $p_{.}(\underline{x}) = p(\underline{x}, 1) + p(\underline{x}, 2) - p(\underline{x}, 1)p(\underline{x}, 2)$. Given only that the two observers detected a total of n different animals on the survey, the appropriate likelihood is just a distance sampling likelihood, similar to Equation (11.2):

$$L_n(N, \underline{\theta}, \underline{\phi}) = \binom{N}{n} (1 - \pi_c E[p])^{N-n} \prod_{i=1}^{n} \pi_c p_{.}(\underline{x}_i) \pi(\underline{x}_i)$$

(12.8)

Because detection probability at distance zero is unknown, however, this likelihood is inadequate for estimation of abundance. The mark-recapture component of the survey provides the additional information that allows estimation of absolute abundance. The observers are the "capture occasions", so if animal i was detected by observer 1, then $\omega_{1i} = 1$, and similarly for observer 2 and ω_{2i}. The conditional likelihood for the parameter vector of the detection function $(\underline{\theta})$, given the n observed capture histories, is just the conditional mark-recapture likelihood Equation (11.22):

$$L_\omega(\underline{\theta}) = \prod_{i=1}^{n} \frac{\prod_{s=1}^{2} p(\underline{x}_i, l_s)^{\omega_{si}} [1 - p(\underline{x}_i, l_s)]^{1-\omega_{si}}}{p_{.}(\underline{x}_i)}$$

(12.9)

where $l_s = s$ indexes observer. The full likelihood for a double-platform distance survey is

$$L(N, \underline{\theta}, \underline{\phi}) = L_n(N, \underline{\theta}, \underline{\phi}) \times L_\omega(\underline{\theta})$$

(12.10)

$$= \binom{N}{n} (1 - \pi_c E[p])^{N-n} \times$$

$$\prod_{i=1}^{n} \left\{ \pi_c \pi(\underline{x}_i) \prod_{s=1}^{2} p(\underline{x}_i, l_s)^{\omega_{si}} [1 - p(\underline{x}_i, l_s)]^{1-\omega_{si}} \right\}$$

This likelihood is easily generalized for the case in which there are survey-level variables other than observer, by defining the vector of survey-level variables on occasion s, $\underline{l}_s$, appropriately.

Notice the similarity between the likelihood Equation (12.10) above, and the mark-recapture likelihood for static populations, Equation (11.29), derived in Section 11.3.7. Aside from the way the state model $\pi(\underline{x})$ is formulated, the two are identical. The double-platform distance sampling model can be thought of as a particular type of two-sample mark-recapture model in which distance x has a special role, and $\pi(x)$ is known.

12.3.2 Conditional likelihood

As was the case with the mark-recapture methods considered in Section 11.3.6, abundance can be estimated from the conditional likelihood Equation (12.9) without an explicit state model, using a Horvitz–Thompson-like estimator:

$$\hat{N} \;=\; \sum_{i=1}^{n} \frac{1}{\pi_c \hat{p}.(\underline{x}_i)} \qquad (12.11)$$

where $\hat{p}.(\underline{x}_i)$ is the MLE of $p.(\underline{x}_i)$, obtained by maximizing Equation (12.9) with respect to $\underline{\theta}$.

Huggins (1989) and Alho (1990) showed that this estimator is asymptotically unbiased, but it can be very biased even for large n when some $p.(\underline{x}_i)$ are small (Borchers, 1996). The behaviour of the estimator can be improved by using knowledge of $\pi(x)$. This is done by using the expected value of $p.(\underline{x})$ with respect to distance, x,

$$\hat{\mu}(\underline{x}_i) \;=\; \int_{0}^{w} \hat{p}.(\underline{x}_i)\pi(x)\,dx \qquad (12.12)$$

in place of $\hat{p}.(\underline{x}_i)$ in the denominator:

$$\hat{N} \;=\; \sum_{i=1}^{n} \frac{1}{\pi_c \hat{\mu}(\underline{x}_i)} \qquad (12.13)$$

Note that

- $\hat{\mu}(\underline{x})$ is the estimated "effective strip half-width" for animals with animal-level variables $\underline{x}$,
- if x is the only animal-level variable, then $\hat{\mu}(\underline{x}) = \hat{\mu}$, and

- the estimator is readily generalized to include survey-level variables $\underline{l}$ other than observer index; we have not done so here for simplicity of presentation.

The estimator of Equation (12.13) has been found to perform better than the estimator of Equation (12.11) in simulation studies (Borchers, 1996); using knowledge of $\pi(x)$ helps!

↑
Technical
section

We illustrate the methods by example below, but before we do so we should point out that there is now a fairly substantial literature on these methods, and that the above is only one of a number of possibilities. Similar methods have been developed by Alpizar-Jara and Pollock (1999), Palka (1995) and Manly *et al.* (1996), among others. Sophisticated double-platform methods that differ somewhat from the above have been developed by Schweder (1990), Schweder *et al.* (1999), Skaug and Schweder (1999), Hiby and Lovell (1998), Cooke (1997, unpublished) and Cooke and Leaper (unpublished). The basis for these methods is much the same as that described above, but they have some features not present in the relatively simple model we present here. These include

An availability model When animals are unavailable for detection for a substantial period of time they are in detectable range, models that incorporate randomness in animal availability may perform substantially better than ones that do not. One can think of "available for detection" as one aspect of an animal's state and extend the state model to include a model for the random process governing this state. In this case, the state model is dynamic and the analysis methods are more complicated. Details of estimation methods can be found in the references above. Minke whales in the North Atlantic provide an example. They are underwater and unavailable for detection for a substantial proportion of the time they are in detectable range. In this case, the dynamic state model for availability is a model for whale surfacing behaviour.

Incorporating uncertainty about duplicates There is substantial uncertainty in some applications, about whether or not a detection is a "duplicate" (a recapture, in mark-recapture terms). The likelihoods above do not accommodate this. Methods do exist for doing so and the interested reader is referred to the papers above for details.

Methods which are more robust to unmodelled heterogeneity in line transect surveys than those described here have been developed by Laake (1999), building on the approach of Palka (1995).

12.3.3   Example: North Sea harbour porpoise

Harbour porpoise abundance in the North Sea was estimated from data collected on the 1994 Small Cetacean Abundance in the North Sea (SCANS) survey. The shipboard component of the survey involved simultaneous double-platform survey with eight vessels, with one vessel per stratum for the most part. The survey involved one observer team searching with binoculars well ahead of the vessel, and the other (the "primary" team) searching with naked eye. The detection function of the primary observer team only is estimated. This design involves a simpler likelihood to estimate the primary team's detection function parameters $\underline{\theta}$ than those presented above. Harbour porpoise occur in groups, and the detection unit was group. Given that a group with animal-level variables $\underline{x}$ was detected by the team searching with binoculars, the probability that the primary team detects it when searching with effort variables $\underline{l}$ is binary, with parameter $p(\underline{x}, \underline{l})$. The likelihood for $\underline{\theta}$ is therefore simply

$$L_{\omega_2}(\underline{\theta}) \;=\; \prod_{i=1}^{n_1} p(\underline{x}_i, \underline{l}_i)^{\omega_{2i}} [1 - p(\underline{x}_i, \underline{l}_i)]^{1-\omega_{2i}} \qquad (12.14)$$

where the platform searching with binoculars detected n_1 groups, $\underline{l}_i$ is the survey-level variables corresponding to the ith group, and ω_{2i} is the capture indicator for the primary team, which is 1 if the primary team detected the ith group, and zero otherwise. A logistic function form was assumed for $p(\underline{x}, \underline{l})$, allowing the likelihood to be maximized using standard logistic regression software. The fitted detection function, averaged over all explanatory variables other than perpendicular distance, is shown in Figure 12.3.

Group size and perpendicular distance were the only animal-level explanatory variables ($\underline{x}$) available, while sea-state, vessel code and swell height were available as survey-level variables ($\underline{l}$). Perpendicular distance was found to be most significant, followed by group size; sea state and vessel code were also found to be significant, but swell height was not.

Animal abundance was estimated using an estimator of the form given in Equation (12.13):

$$\hat{N} \;=\; \sum_{i=1}^{n_2} \frac{z_i}{\pi_c \hat{\mu}(\underline{x}_i, \underline{l}_i)} \qquad (12.15)$$

where n_2 is the number of groups detected by the primary team and $\hat{\mu}(\underline{x}_i, \underline{l}_i)$ is the estimated effective strip half-width corresponding to group i for the primary team.

Confidence intervals were estimated using the percentile method with a nonparametric bootstrap based on transects, resampling transects separately within vessel and stratum. Details of methods and the application are contained in Borchers et al. (1998b,a).

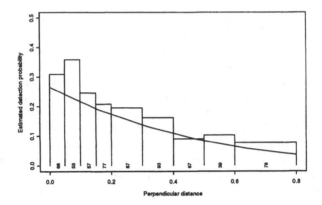

Figure 12.3. Estimated detection probability for harbour porpoise from the primary platform, shown as a function of perpendicular distance only. The histogram shows the observed proportion of trial sightings that were detected by the primary platform. Heterogeneity makes the histogram a positively biased estimate of detection probability; the fitted curve models the heterogeneity. The number of trial sightings in each histogram bar is shown at the bottom of the bar.

12.4   Double-platform migration surveys

In all migration surveys animals passing outside the watch periods are missed. In most migration surveys, an unknown proportion of the animals passing within the watch periods are also missed. When this occurs, mark-recapture methods can be used within watch periods to estimate the proportion missed. This is done by having two teams survey at the same time, independently of one another, and recording data that allow the analyst to identify which animals were detected by both teams. The method is very similar in concept to double-platform distance sampling surveys, except that (a) time takes the place of distance and (b) the temporal state model is not known. Conceptually each team is a "capture occasion" and animals that are detected by both teams are "recaptures" (again called "duplicates" or "duplicate detections" in this context). We will call this sort of survey a double-platform migration survey.

Current practice is to model n_j, the total of these counts between the two platforms, as described for the method with just one observer team (Section 10.2). A separate mark-recapture analysis is then carried out, to estimate the number of animals passing during watch periods that are missed, from which a multiplicative correction factor is estimated. Account needs to be taken of the additional variance due to estimation of the number passing in watch periods (see Buckland et al., 1993). To reduce bias due to unmodelled heterogeneity in the detection probabilities between animals, covariates relating to the environment, the detected animals and the observers can be recorded and included in the model.

A more integrated approach is possible, which is analytically just a version of the mark-recapture method described in Section 11.3.7. To see this, it helps to distinguish time from survey occasion. With the mark-recapture surveys considered so far, the survey occasions occur at different times and the whole population is surveyed on all occasions. With double-platform migration surveys, the survey occasions occur at the same time and place, but only part of the population is surveyed. Think of the animals as a queue, with lots of animals in the middle (at the migration peak) and few at the front (migration start) and back (migration end). This queue files past the observers and pieces of it that pass in watch periods get sampled, but you can think of the queue as static, and the watch periods as being locations along the queue that are sampled.

The location of an animal in this queue is an animal-level variable and there is a state model that determines where the animal falls in the queue, i.e. when it passes the observation point. Suppose animal i passes at time x_i and let $\pi(x_i)$ be the probability density of the time it passes. This describes the shape of the migration curve, with higher density in the middle than at the ends of the migration period. We assume that the x_is are independent, identically distributed random variables ($i = 1, \ldots, N$). Conceptually, each animal is allocated its x_i once for the migration, hence the state model is a static one. Now in general there is a vector of animal-level variables that affect animals' detectability. As usual, we call it $\underline{x}$. If we include the time of passing x in this vector, then the methods and likelihoods of Section 11.3.7 apply directly. In addition to providing estimates of the numbers of animals missed in the migration periods, maximizing the likelihood Equation (11.29) (with $S = 2$ as there are two observer teams) will provide an estimate of the migration curve $\pi(x)$ and the total number of animals in the migration, N.

Note that capture history will usually not affect detectability since animals are not physically captured, so behavioural response will not occur and detection probability will not depend on $\underline{\omega}_{si}$. Captured animals might be more detectable than uncaptured animals on average, but in a migration survey it is unlikely that this will be because of animals physically reacting to "capture".

Technical
section
↓

To illustrate how a double-platform migration survey is analytically a sort of hybrid mark-recapture and migration count method, consider the simplest case, in which detection probability is constant but unknown. The relevant likelihood is the mark-recapture likelihood Equation (11.29) with $S = 2$, $\underline{x} = x$ and constant $p(x) = p$ for x within watch periods, and $p(x) = 0$ for x outside watch periods:

$$L(N, p, \underline{\phi}) = \binom{N}{n} E\left[(1-p(x))^2\right]^{N-n} \times$$

$$\prod_{i=1}^{n}\left\{\pi(x_i)\prod_{s-1}^{2}p^{\omega_{si}}(1-p)^{1-\omega_{si}}\right\}$$

$$= \binom{N}{n}[1-p.\pi_c]^{N-n}\prod_{i=1}^{n}\pi(x_i)$$

$$\times\prod_{i=1}^{n}\prod_{s-1}^{2}p^{\omega_{si}}(1-p)^{1-\omega_{si}} \qquad (12.16)$$

since

$$E\left[(1-p(x))^2\right] = 1 - \int_0^T (2p(x)-p(x)^2)\pi(x)\,dx$$

$$= 1 - \int_0^T p.(x)\pi(x)\,dx$$

$$= 1 - p.\pi_c \qquad (12.17)$$

where the migration starts at $x = 0$ and ends at $x = T$, $p.(x) = p. = 2p + p^2$ is the probability of at least one of the two observer teams detecting an animal passing within a watch period ($p.(x) = 0$ outside the covered period), and π_c is the integral of the migration curve $\pi(x)$ over the covered period, as before.

Notice

- that this mark-recapture likelihood reduces to the simple migration count likelihood for unbinned data, Equation (10.6), if $p = 1$ (in which case all the ω_{si} are 1), and

- the similarity to the double-platform distance sampling likelihood Equation (12.10); analytically the two methods are identical except for the special role of distance x and the fact that $\pi(x)$ is known in distance sampling likelihoods.

↑
Techni
section

Other extensions are often introduced in the form of multiplicative correction factors. These include a correction for night passage rate, where this differs from the rate during watch periods, and a correction to the group size estimate, for when animals migrate in groups whose sizes cannot be accurately counted. When such corrections are estimated from additional (independent) data, models can be specified for the data, from which likelihoods can be developed. These likelihoods can then be multiplied by the likelihood for the migration counts, if an integrated modelling approach is required.

12.5   Other possibilities

The mark-recapture likelihoods developed in Chapter 11 provide a basis for simultaneous estimation of abundance and spatial distribution (by including an appropriate $\pi(\underline{x}_s)$ with a spatial component). Sections 12.3 and 12.4 above involved static populations: $\underline{x}$ was assumed not to change between capture occasions – because the occasions occurred effectively simultaneously. Neither involved an unknown spatial state model, but this could be added to the likelihood functions without difficulty (although estimation may then be difficult). In the case of a double-platform distance sampling survey, a static spatial state model might be appropriate. In the case of a double-platform migration survey, some form of time-varying spatial state model might be most appropriate. Consider, for example, a double-platform migration survey in which distance from shore was a relevant spatial variable. If the distribution of animals from shore remained constant throughout the survey, a static state model might be appropriate, whereas if the distribution shifted, a spatial state model with population-level drift might be appropriate.

In most mark-recapture applications, a static spatial state model will be inappropriate; this would not be the case for plants, but animals tend to move about. Full likelihood models for mark-recapture surveys of the sort developed in Section 11.3.7 have not appeared in the literature to date, but they would seem to provide a basis for developing a wide variety of mark-recapture models. These models also provide a basis for spatial modelling with removal methods, and for combining removal methods with distance sampling, mark-recapture, or other methods.

12.6   Summary

We have covered three methods that integrate apparently different methods:

- A generalization of spatial modelling methods to accommodate uncertain detection. Integration of distance sampling methods of Chapters 7 and 11 with spatial modelling methods of Chapter 10.

- Integration of distance sampling methods of Chapters 7 and 11 with mark-recapture methods of Chapter 11.

- Integration of migration survey methods of Chapter 10 with mark-recapture methods of Chapter 11.

These are only a few of many possibilities. The likelihood approach we have adopted gives a conceptual framework that makes it easier to see links between what at first glance appear to be unrelated methods. The similarity between migration counts of whales and surveys of pelagic fish eggs is a case in point. We expect that in future there will be increasing use and development of methods and models that cross what were previously considered methodological boundaries. We have found the likelihood framework useful in this context.

12.7   Exercises

Exercise 12.1   Show that when when $\pi(x) = 1/A$ (a spatial model with uniformly distributed animals), the MLE of abundance from the spatial model likelihood is equivalent to the conventional distance sampling estimator of abundance.

Exercise 12.2   Show that Horvitz–Thompson-like estimator gives a conventional distance estimator when detection probability depends only on x.

Exercise 12.3   Equation (12.5) gives the pdf of the distance to the next detection, d, given w and (u_0, v_0). From this, derive an expression for the pdf for the case in which density varies linearly across the width of the strip, i.e. $D(d) = \alpha + \beta x$, where x is distance perpendicular to the line, zero on the line, and $-w \leq x \leq w$.

13
Dynamic and open population models

Key idea: model population dynamics to improve estimates of time-varying abundance.

Number of surveys:	Multiple surveys are needed to model abundance of open populations.
State model:	May include elements for any or all of mortality, recruitment, movement, species interactions, mating and genetic diversity.
Observation models:	Depend on the method(s) of surveying the population. These methods may be any of the single-survey methods in this book. The approach also provides a framework for extending the multiple-survey methods to the open population case.

Likelihood function and key notation:

$$L(\underline{N}_1; \underline{\phi}; \underline{\theta}) \; = \; \sum_{\boldsymbol{N} \to \boldsymbol{y}} \prod_{s=1}^{S} L_s(\underline{N}_s; \underline{\phi}; \underline{\theta})$$

$\underline{N}_s$: cector of animal abundance on occasion s $(s = 1, ..., S)$, where the elements correspond to age and other categories of animal.

$\boldsymbol{N}$: matrix of animal abundance comprising the set of S vectors $\underline{N}_s$.

$\underline{y}_s$: vector of observations on occasion s, containing counts of animals in each category, $\underline{n}_s$ (see text).

$\boldsymbol{y}$: matrix of observations comprising the set of S vectors $\underline{y}_s$.

$\sum_{\boldsymbol{N} \to \boldsymbol{y}}$: the sum over all states $\boldsymbol{N}$ that could give rise to $\boldsymbol{y}$, from a population of $\underline{N}_1$ animals on occasion 1.

$\underline{\phi}$: the parameters of the underlying state model.

$\underline{\theta}$: the parameters of the observation model(s).

$L_s(\underline{N}_s; \underline{\phi}; \underline{\theta})$: the likelihood corresponding to occasion s, $s = 1, ..., S$. This is determined by the type of survey conducted on occasion s.

13.1 Introduction

We concentrate on closed populations in this book and provide only a brief introduction to open population models for two reasons. First, adequate coverage of integrated modelling for open populations requires a book in its own right. Second, the tools to fit such integrated models have only recently been developed, and more research is needed to explore their potential.

We provide a general framework for combining state models with observation models, which provides the means to estimate output parameters such as abundance.[1] The observation models allow us to calibrate or fit the state models to observed data. We define the population dynamics model by setting up a state equation, and the population components at any point in time are mapped to observed data, or to estimates of population parameters, through an observation equation.

---

[1]Because this chapter is almost all technical, we do not explicitly mark technical sections.

The methods are applicable to dynamic closed populations as well as open populations. We start by developing examples with simple dynamic closed populations (ones with a spatial state model and movement), and then move on to open populations.

13.2 Dynamic closed population example

We formulate a fairly simple closed-population example both as we have thus far in the book, and as a state-space model, to clarify how the sophisticated models and methods used for open population models are natural extensions of the models and methods for closed populations. We do not aim at realism in this section; instead, we choose simple models to illustrate the ideas (see Exercise 13.3 for a still simpler closed population example). We frame the example in terms of the grassland-woodland scenario we used to illustrate state models in Chapter 9, with a Markov movement model.

13.2.1 State model

The simplest possible spatial state model describes the members of the population in terms of only two states: in grassland ($x = 0$) or in woodland ($x = 1$). In this case the state of the population on occasion s could be described by the simple population vector $N_s = (N_s(0), N_s(1))^T$, where $N_s(x)$ is the number of animals in habitat x on occasion s. Suppose we did a plot survey of the population, with 26% coverage of each habitat, as shown in Figure 13.1. This splits the population in each habitat into two states: those in the covered region (denoted by $y = 1$), and those outside it (denoted $y = 0$). The state of the population on occasion s can now be described in terms of the state vector (x_s, y_s) as follows

$$\underline{N}_s \;=\; \begin{pmatrix} N_s(0,1) \\ N_s(1,1) \\ N_s(0,0) \\ N_s(1,0) \end{pmatrix} \tag{13.1}$$

$$=\; \begin{pmatrix} \text{number in grassland inside covered region on occasion } s \\ \text{number in woodland inside covered region on occasion } s \\ \text{number in grassland outside covered region on occasion } s \\ \text{number in woodland outside covered region on occasion } s \end{pmatrix}$$

To keep the model relatively simple, we assume that survey occasions are equally spaced in time; this avoids the need to take account of times between surveys in modelling movement. We also assume that animals are distributed uniformly and independently within each habitat, so that the

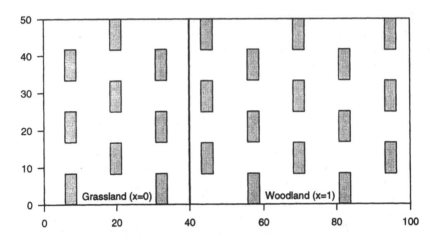

Figure 13.1. Location of plots in a plot survey of a population that moves between grassland and woodland.

probability of an animal in habitat x being inside the covered region is $\pi_c(x) = a(x)/A(x)$, where $a(x)$ is the area of the covered region in habitat x and $A(x)$ is the total area of habitat x. If we let $y_s = 1$ when an animal is in the covered region on occasion s (and zero otherwise), and x_s index its habitat on occasion s, we can write the probability mass function of y_s, given x_s as

$$\pi(y_s \,|\, x_s) \;\; = \;\; \pi_c(x_s)^{y_s}(1 - \pi_c(x_s))^{1-y_s} \qquad (13.2)$$

In the example above, $a(0) = 520$, $A(0) = 2,000$, $a(1) = 780$ and $A(1) = 3,000$, so that $\pi_c(0) = \pi_c(1) = 26\%$.

At the individual level, we assume a (Markov) movement model such that an animal in habitat type x on occasion $s-1$ is also in this habitat on occasion s with a probability ϕ_x. We can write the probability mass function of x_s (the animal's habitat on occasion s), given x_{s-1} (the animal's habitat on occasion $s-1$), as follows:

$$\begin{aligned} \pi(x_s|0) &= (1 - \phi_0)^{x_s}\phi_0^{1-x_s} \\ \pi(x_s|1) &= \phi_1^{x_s}(1 - \phi_1)^{1-x_s} \end{aligned} \qquad (13.3)$$

where $\pi(x_s|0)$ is shorthand for $\pi(x_s|x_{s-1} = 0)$ and $\pi(x_s|1)$ is shorthand for $\pi(x_s|x_{s-1} = 1)$.

We assume animals move independently of one another. This, together with our assumption of a uniform, independent state model above, implies that the number of animals in state (x_s, y_s) on occasion s, given that they were in state (x_{s-1}, y_{s-1}) on occasion $s-1$, is a binomial random variable with parameters $N_{s-1}(x_{s-1}, y_{s-1})$ and $\pi_c(x_s)\pi(x_s|x_{s-1})$. It follows that the probability mass function of $N_s(x_s, y_s)$, given $\underline{N}_{s-1}$, is the sum of four independent binomial random variables, one for each of the four possible states (x_{s-1}, y_{s-1}) on occasion $s - 1$.

We write the probability that the population is in state $\underline{N}_s$ on occasion s, given that it was in state $\underline{N}_{s-1}$ on occasion $s-1$, as $P(\underline{N}_s|\underline{N}_{s-1})$. This is our state model; it describes the spatial distribution and Markov dynamics of the movement process (how the state of the population changes with time).

We can specify the state model succinctly in matrix terms as follows:

$$\underline{N}_s \;\; = \;\; DM\underline{N}_{s-1} + \underline{\varepsilon}_s \tag{13.4}$$

where

$$M \;\; = \;\; \left[\begin{array}{cccc} \phi_0 & (1 - \phi_1) & \phi_0 & (1 - \phi_1) \\ (1 - \phi_0) & \phi_1 & (1 - \phi_0) & \phi_1 \end{array} \right] \tag{13.5}$$

is the movement matrix, and

$$D \;\; = \;\; \left[\begin{array}{cc} \pi_c(0) & 0 \\ 0 & \pi_c(1) \\ (1 - \pi_c(0)) & 0 \\ 0 & (1 - \pi_c(1)) \end{array} \right] \tag{13.6}$$

is the spatial distribution matrix, and $\underline{\varepsilon}_s$ is a 4×1 vector of error terms with mean zero (see Exercise 13.1).

This describes the states of the population, and how animals change from one state to another. We also need an observation model, to define how the states of the population map to our observations.

13.2.2 Observation model

The observation model describes the probability of the observed outcome on occasion s, given the state of the population on occasion s (i.e. $\underline{N}_s$): we denote it $\underline{y}_s$. For example, if we detect $n_s(x)$ animals in habitat x on occasion s, the outcome is

$$\underline{y}_s \;\; = \;\; \underline{n}_s = \left[\begin{array}{c} n_s(0) \\ n_s(1) \end{array} \right] \tag{13.7}$$

In the case of a plot survey, we detect all the animals in the covered region, so given the state of the population, there is no randomness in the outcome;

all that we need to do is to tally up the numbers of animals in the covered region. We can do this in matrix terms using a matrix $\boldsymbol{B}$, as follows:

$$\underline{n}_s = \boldsymbol{B}\underline{N}_s \tag{13.8}$$

where

$$\boldsymbol{B} = \begin{bmatrix} 1 & 0 & 0 & 0 \\ 0 & 1 & 0 & 0 \end{bmatrix} \tag{13.9}$$

13.2.3 Likelihood

Multiplying the state and observation models for occasion s together, we get the probability that the population is in state $\underline{N}_s$ and we detect $\underline{n}_s$, given the state of the population on occasion $(s-1)$:

$$P(\underline{N}_s, \underline{n}_s \mid \underline{N}_{s-1}) = P(\underline{N}_s \mid \underline{N}_{s-1}) \times P(\underline{n}_s \mid \underline{N}_s) \tag{13.10}$$

(Note that $P(\underline{n}_s \mid \underline{N}_s)$ is degenerate: it is one if $N_s(0,1) = n_s(0)$ and $N_s(1,1) = n_s(1)$; it is zero otherwise. With surveys other than plot sampling, it is a proper probability function.)

For static populations, once $\underline{N}_{s-1}$ is given, there is no uncertainty about $\underline{N}_s$: if an animal starts in habitat x, it stays in habitat x. All uncertainty about $\underline{N}_s$ stems from uncertainty about the initial state of the population, $\underline{N}_1$. For dynamic models, this is no longer the case. Even if we knew the initial state of the population, $\underline{N}_1$, there would be uncertainty about its state on later occasions, $\underline{N}_s$ $(s > 1)$, because an animal initially in habitat x_1 might have moved out of this habitat by occasion s.

The probability that the population passes through states $\underline{N}_1$, $\underline{N}_2$, to $\underline{N}_S$, and we observe $\underline{n}_1$, $\underline{n}_2$, to $\underline{n}_S$, can be written as

$$P(\boldsymbol{N}, \boldsymbol{n}) = \prod_{s=1}^{S} P(\underline{N}_s, \underline{n}_s \mid \underline{N}_{s-1}) \tag{13.11}$$

where $\boldsymbol{N}$ is the matrix with columns $\underline{N}_1, \underline{N}_2, \ldots, \underline{N}_S$, and $\boldsymbol{n}$ is the matrix with columns $\underline{n}_1, \underline{n}_2, \ldots, \underline{n}_S$.

Note that $P(\boldsymbol{N}, \boldsymbol{n})$ depends on the model parameters, although we do not write these out explicitly here. What are the model parameters? There are no observation model parameters because all animals in the covered region are detected. There are two movement parameters, ϕ_0, ϕ_1, and we parametrize the model further with a vector of initial abundance parameters for woodland and grassland:[2] $\underline{N}_1 = (N_1(0), N_1(1))^T$. There are also two parameters associated with animal distribution ($\pi_c(0)$ and $\pi_c(1)$) but these are known, by design.

Because we don't observe N, Equation (13.11) can't be used directly for inference; we need a likelihood for the unknown parameters given what was observed. To get this, we need to sum Equation (13.11) over all possible states N of a population of N animals that could have given rise to the data we observed ($\boldsymbol{n}$):

$$L(\underline{N}_1, \underline{\phi} \,|\, \boldsymbol{n}) \;=\; \sum_{N \to n} P(\boldsymbol{N}, \boldsymbol{n}) \tag{13.12}$$

where $\boldsymbol{N} \to \boldsymbol{n}$ indicates the sum over all states $\boldsymbol{N}$ of a population of initially $\underline{N}_1$ animals that could have given rise to $\boldsymbol{n}$.

Even with our simple example, this likelihood is more complicated than those encountered for closed static populations and it is more difficult to evaluate and maximize. There are specialized methods that take advantage of the sequential nature of the likelihood for estimation. We mention these briefly below.

13.3   Open population example

Suppose we have a single population for which we wish to model only three age categories of animal: 0, 1 and 2 years or older. We will assume a juvenile (age 0) survival rate of ψ_j in year 0 and a constant adult (age ≥ 1 years) survival rate of ψ_a in subsequent years. The number of offspring per year per female of age ≥ 2 is assumed to be Poisson with rate λ, independent of the number of males, and the probability that a young animal is male is assumed to be α. There is no known take (such as a harvest or cull) of animals. Mortality occurs largely in winter, and young are born in summer.

We further assume that the above population is surveyed annually, using line transect sampling with transects that cover the same fraction π_c of the survey area every year. Suppose that three categories of animal can readily be distinguished in the field: males aged ≥ 1, females aged ≥ 1, and first-year juveniles, so that we have an abundance estimate (and corresponding variance estimate) for each category. We suppose also that these surveys

---

[2]We could also parametrize it in other ways. For example, with one abundance parameter, N, representing initial abundance in both habitats combined, and a parameter specifying the probability that an animal is in habitat 0 on the first occasion: $\pi_0 = \text{prob}\{x_1 = 0\}$.

are conducted in spring, after winter mortality but before the new young are born. Finally, we assume that animals are distributed uniformly in the survey region, independently of one another and independently of their previous locations.

13.3.1 State model

The order of terms in the state equation depends on the chronology within a year of births, mortality and surveys. If mortality occurs throughout the year, it may be necessary to separate it out into components (for example, mortality between breeding season and the survey, and between the survey and the breeding season). Similarly, births might occur throughout the year. For our example, the state model may be expressed as follows:

$$E[\underline{N}_s] = \mathbf{DSRI}\underline{N}_{s-1} \tag{13.13}$$

where

s denotes year s, $s = 1, ..., S$;

$\underline{N}_s$ is a vector of length q that corresponds to numbers of animals in each of q categories present in year s, after natural mortality but before new births;

$E[\underline{N}_s]$ is the expectation of $\underline{N}_s$ conditional on $\underline{N}_{s-1}$;

$\mathbf{I}$ is a $q \times q$ matrix that increments ages by one year;

$\mathbf{R}$ is a $q \times q$ matrix that models new births;

$\mathbf{S}$ is a $q \times q$ diagonal matrix, whose diagonal elements are the annual survival probabilities of the q categories of animal in year s;

$\mathbf{D}$ is a $q \times q$ matrix governing animal distribution between the covered and uncovered regions.

In general, we would allow survival rates to vary by age and sex, and birth rates to vary by age. While in principle each category in $\underline{N}_s$ might have its own death rate and, for females, birth rate, in practice any plausible data set will be largely uninformative for many of these parameters. Some degree of parsimony is therefore required. For example, death rates and birth rates might be assumed to be some smooth function of age and of animal density (to allow for density-dependent effects). This function might be parametrized (Trenkel *et al.*, 2000), or partially specified, using semi-parametric smooths (Wood, 2001). One or more additional parameters might be introduced to allow for different survival rates between males and females, and rates might also be allowed to be a function of year, habitat, weather or other relevant covariates. If there is a harvest, cull or introduction, these animals are subtracted or added at the appropriate time of year.

Stochasticity could be modelled by making a simple assumption, such as multivariate normality, about the distribution of $\underline{N}_s$, given the terms on the right-hand side of the above equation. Simulation-based inference methods (below) allow models to be fitted in which the stochasticity arises more naturally, through processes built into $\mathbf{D}$, $\mathbf{R}$ and $\mathbf{S}$. In our simple example, these processes are: the number of animals in the covered region on any occasion is binomial with parameters N and π_c (the probability of being in the covered region); Poisson births with rate λ for each female aged ≥ 2; given a number of births $N_0 > 0$, the number of male births N_{0m} is binomial with parameters N_0 and α; the number of these N_0 animals that survive to age one (N_1) is binomial with parameters N_0 and ψ_j; and the number of animals of age i, N_i, that survive to age $i + 1$, $i \geq 1$, is binomial with parameters N_i and ψ_a.

For our example, there are two sexes, three age classes, and two locations (in or out of the covered region), so that $q = 2 \times 3 \times 2 = 12$ and the state model is :

$$E[\underline{N}_s] = \mathbf{DSRI}\underline{N}_{s-1} \tag{13.14}$$

where

$$\underline{N}_s = \begin{bmatrix} N_{c0ms} \\ N_{c1ms} \\ N_{c2ms} \\ N_{c0fs} \\ N_{c1fs} \\ N_{c2fs} \\ N_{u0ms} \\ N_{u1ms} \\ N_{u2ms} \\ N_{u0fs} \\ N_{u1fs} \\ N_{u2fs} \end{bmatrix} \tag{13.15}$$

with the first suffix indicating location (c for those in the covered region, u for those not), the second indicating age (0, 1 or ≥ 2), the third indicating sex, and the fourth year. The matrices $\mathbf{S}$, $\mathbf{R}$ and $\mathbf{I}$ are

$$\mathbf{S} = \begin{bmatrix} \mathbf{S}^* & 0 \\ 0 & \mathbf{S}^* \end{bmatrix} \quad \mathbf{R} = \begin{bmatrix} \mathbf{R}^* & 0 \\ 0 & \mathbf{R}^* \end{bmatrix} \quad \mathbf{I} = \begin{bmatrix} \mathbf{I}^* & 0 \\ 0 & \mathbf{I}^* \end{bmatrix}$$

where

$$I^* = \begin{bmatrix} 0 & 0 & 0 & 0 & 0 & 0 \\ 1 & 0 & 0 & 0 & 0 & 0 \\ 0 & 1 & 1 & 0 & 0 & 0 \\ 0 & 0 & 0 & 0 & 0 & 0 \\ 0 & 0 & 0 & 1 & 0 & 0 \\ 0 & 0 & 0 & 0 & 1 & 1 \end{bmatrix} \qquad (13.16)$$

$$R^* = \begin{bmatrix} 0 & 0 & 0 & 0 & 0 & \alpha\lambda \\ 0 & 1 & 0 & 0 & 0 & 0 \\ 0 & 0 & 1 & 0 & 0 & 0 \\ 0 & 0 & 0 & 0 & 0 & (1-\alpha)\lambda \\ 0 & 0 & 0 & 0 & 1 & 0 \\ 0 & 0 & 0 & 0 & 0 & 1 \end{bmatrix} \qquad (13.17)$$

$$S^* = \begin{bmatrix} \psi_j & 0 & 0 & 0 & 0 & 0 \\ 0 & \psi_a & 0 & 0 & 0 & 0 \\ 0 & 0 & \psi_a & 0 & 0 & 0 \\ 0 & 0 & 0 & \psi_j & 0 & 0 \\ 0 & 0 & 0 & 0 & \psi_a & 0 \\ 0 & 0 & 0 & 0 & 0 & \psi_a \end{bmatrix} \qquad (13.18)$$

while the matrix D is

$$D = \begin{bmatrix} D_c & D_c \\ D_u & D_u \end{bmatrix} \qquad (13.19)$$

where

$$D_c = \begin{bmatrix} \pi_c & 0 & 0 & 0 & 0 & 0 \\ 0 & \pi_c & 0 & 0 & 0 & 0 \\ 0 & 0 & \pi_c & 0 & 0 & 0 \\ 0 & 0 & 0 & \pi_c & 0 & 0 \\ 0 & 0 & 0 & 0 & \pi_c & 0 \\ 0 & 0 & 0 & 0 & 0 & \pi_c \end{bmatrix} \qquad (13.20)$$

and

$$D_u = \begin{bmatrix} (1-\pi_c) & 0 & 0 & 0 & 0 & 0 \\ 0 & (1-\pi_c) & 0 & 0 & 0 & 0 \\ 0 & 0 & (1-\pi_c) & 0 & 0 & 0 \\ 0 & 0 & 0 & (1-\pi_c) & 0 & 0 \\ 0 & 0 & 0 & 0 & (1-\pi_c) & 0 \\ 0 & 0 & 0 & 0 & 0 & (1-\pi_c) \end{bmatrix}$$

$$(13.21)$$

In fact, Equation (13.14) is only exact if the model is linear in the states. However, the Bayesian fitting algorithms of Section 13.4 are applicable whether the model is linear or not.

13.3.2 Observation model

The observation equation maps the observations for year s to the population categories in $\underline{N}_s$. In our example, line transect sampling (Chapter 7) was used to generate annual population abundance estimates by three categories: males aged ≥ 1, females aged ≥ 1 and juveniles. The data comprise counts of each of these three categories: n_{ms}, n_{fs} and n_{0s} are the numbers of males aged ≥ 1 detected in year s, the number of females aged ≥ 1 detected in year s, and the number of juveniles detected in year s, respectively. We suppose for simplicity that the covered region has area a on every line transect survey.

We need to consider how the three observed counts n_{ms}, n_{fs} and n_{0s} map to the 12 elements in $\underline{N}_s$. If we had conducted complete counts on the population in year s, we would know the total numbers of males, females and juveniles (N_{ms}, N_{fs} and N_{0s}), and the mapping in the observation equation would be straightforward:

$$
\begin{bmatrix} N_{0s} \\ N_{ms} \\ N_{fs} \end{bmatrix} = \begin{bmatrix} \mathbf{B}^* & \mathbf{B}^* \end{bmatrix} \begin{bmatrix} N_{c0ms} \\ N_{c1ms} \\ N_{c2ms} \\ N_{c0fs} \\ N_{c1fs} \\ N_{c2fs} \\ N_{u0ms} \\ N_{u1ms} \\ N_{u2ms} \\ N_{u0fs} \\ N_{u1fs} \\ N_{u2fs} \end{bmatrix} = \mathbf{B}\underline{N}_s \qquad (13.22)
$$

where

$$
\mathbf{B}^* = \begin{bmatrix} 1 & 0 & 0 & 1 & 0 & 0 \\ 0 & 1 & 1 & 0 & 0 & 0 \\ 0 & 0 & 0 & 0 & 1 & 1 \end{bmatrix} \qquad (13.23)
$$

and $\mathbf{B}$ is the aggregation matrix. (Note that this formulation assumes that animals present in the survey correspond to those tallied in $\underline{N}_s$, which is true in our example, for which surveys are conducted after winter mortality, and $\underline{N}_s$ represents numbers by category prior to new births.)

Had the three observed counts n_{ms}, n_{fs} and n_{0s} been from a plot sample, in which all animals in the covered region were detected and none outside it were detected, $\mathbf{B}$ would be $[\mathbf{B}^* \ \mathbf{0}]$, where $\mathbf{0}$ is a 3×6 matrix of zeros.

In our example, the three observed counts n_{ms}, n_{fs} and n_{0s} come from a line transect survey, in which animals in the covered region were detected independently with probabilities (μ_{0s}/w), (μ_{ms}/w) and (μ_{fs}/w) for juveniles, males and females, respectively, on occasion s; no animals outside the covered region can be detected. Here μ_{0s}, μ_{ms} and μ_{fs} are the effective strip half-widths for each category in year s, and w is the strip half-width (the same on all surveys). In this case, the observation equation becomes

$$\underline{y}_s = \mathbf{B}\underline{N}_s + \underline{\varepsilon}_s \tag{13.24}$$

where $\underline{y}_s = (n_{0s}, n_{ms}, n_{fs})^T$ is the vector of observed counts in year s; $\underline{N}_s$ is as before; $\mathbf{B} = [\mathbf{B}^* \ \mathbf{0}]$ is the aggregation matrix, with

$$\mathbf{B}^* = \begin{bmatrix} \frac{\mu_{0s}}{w} & 0 & 0 & \frac{\mu_{0s}}{w} & 0 & 0 \\ 0 & \frac{\mu_{ms}}{w} & \frac{\mu_{ms}}{w} & 0 & 0 & 0 \\ 0 & 0 & 0 & 0 & \frac{\mu_{fs}}{w} & \frac{\mu_{fs}}{w} \end{bmatrix} \tag{13.25}$$

and $\underline{\varepsilon}_s$ is the vector of sampling errors corresponding to $\underline{y}_s$. Note the need now to introduce a vector of errors with expectation zero (since $E[\underline{y}_s] = \mathbf{B}\underline{N}_s$) and variance $Var[\underline{\varepsilon}_s] = Var[\underline{y}_s]$, which, assuming independent counts of the three observed categories of males, females and juveniles, is a diagonal matrix with diagonal elements equal to the variances of the three counts, n_{0s}, n_{ms} and n_{fs}.

Our example has one further complication: the effective strip half-widths μ_{0s}, μ_{ms} and μ_{fs} must be estimated. The integrated modelling option is to have them in the observation equation as unknown parameters. They may then be estimated along with the parameters of the state model; note that in this case, the likelihood corresponding to the observations in year s will be a function both of the counts in year s (n_{0s}, n_{ms} and n_{fs}) and of the perpendicular distances of detected animals from the line, given in Equation (7.10).

A simpler option is to estimate the μs using the conditional maximum likelihood approach of Section 7.2.2, and use the estimates in the aggregation matrix, replacing μ_{0s}, μ_{ms} and μ_{fs} by the estimators $\hat{\mu}_{0s}$, $\hat{\mu}_{ms}$ and $\hat{\mu}_{fs}$ in $\mathbf{B}^*$ above. However, we cannot then assume that $Var[\underline{\varepsilon}_s] = Var[\underline{y}_s]$. This may be circumvented by multiplying both sides of the equation by a diagonal matrix with diagonal elements $w/\hat{\mu}_{0s}$, $w/\hat{\mu}_{ms}$ and $w/\hat{\mu}_{fs}$, which yields

$$\underline{y}_s^* = \mathbf{B}\underline{N}_s + \underline{\varepsilon}_s \qquad (13.26)$$

where

$$\underline{y}_s^* = \begin{bmatrix} \hat{N}_{c0s} \\ \hat{N}_{cms} \\ \hat{N}_{cfs} \end{bmatrix} \qquad (13.27)$$

is the vector of estimated numbers of juveniles, males and females in year s **in the covered region** (see Exercise 13.5); $\underline{N}_s$ is as before; $\underline{\varepsilon}_s$ is the vector of sampling errors corresponding to $\underline{y}_s^*$; $\mathbf{B} = [\mathbf{B}^* \ \mathbf{0}]$ is the aggregation matrix, with

$$\mathbf{B}^* = \begin{bmatrix} 1 & 0 & 0 & 1 & 0 & 0 \\ 0 & 1 & 1 & 0 & 0 & 0 \\ 0 & 0 & 0 & 0 & 1 & 1 \end{bmatrix} \qquad (13.28)$$

We now estimate $Var\,[\underline{\varepsilon}_s]$ by the estimated variance-covariance matrix of the estimated numbers within the covered region, obtained from the conventional line transect analysis. If independent analyses are carried out for each category, then this matrix will be diagonal, with diagonal elements equal to the estimated variances $\widehat{Var}\left[\hat{N}_{c0s}\right]$, $\widehat{Var}\left[\hat{N}_{cms}\right]$ and $\widehat{Var}\left[\hat{N}_{cfs}\right]$. If the analyses are not independent, for example if the effective strip half-width is estimated from data pooled across categories, transect lines can be resampled using the nonparametric bootstrap (Section 7.2.5), and the three abundance estimates evaluated for each resample. The sample variances and covariances of these estimates then provide the estimated variance-covariance matrix.

13.3.3 Likelihood

We can express the full likelihood as follows:

$$L(\underline{N}_1; \underline{\phi}; \underline{\theta}) = \sum_{N \to y} \prod_{s=1}^{S} L_s(\underline{N}_s; \underline{\phi}; \underline{\theta}) \qquad (13.29)$$

where

$\underline{N}_s$ is the vector of animal abundance on occasion s $(s = 1, ..., S)$, where the elements correspond to the age and sex categories of animal (up to some maximum age);

N is the matrix of animal abundance comprising the set of S vectors $\underline{N}_s$;

y is the matrix of observations comprising the set of S vectors $\underline{y}_s$;

$\underline{\phi}$ is the vector of parameters of the underlying state model;

$\underline{\theta}$ is the vector of parameters of the observation model;

$L_s(\underline{N}_s; \underline{\phi}; \underline{\theta})$ is the likelihood corresponding to the data recorded and the states on occasion s, given the states at time point $s-1$, $s = 1, ..., S$;

$\sum_{N \to y}$ is the sum over all states N that could give rise to y, from a population of $\underline{N}_1$ animals on occasion 1.

For our example, if we adopt the conditional method, $\underline{y}_s$ is the vector of estimated abundances in the covered region on occasion s ($s = 1, ..., S$), where the elements correspond to observable categories of animal; for the integrated method, it is the corresponding vector of counts, together with the distances of detected animals from the line.

The likelihood on occasion s has two components. The first reflects demographic stochasticity, and is the distribution of $\underline{N}_s$ given $\underline{N}_{s-1}$. The second reflects stochastic error in the observations, and is the distribution of $\underline{y}_s$ given $\underline{N}_s$. Because the states are unobserved, we must integrate (or more strictly sum) over them, which we do using simulated inference (below).

The form of the likelihood is determined by the survey method used. In our example, the parameters $\underline{\theta}$ are the parameters of the detection function. The effective strip half-width μ is a function of these parameters (Section 7.2.2). If the conditional method is used for fitting $\underline{\theta}$, then the likelihood corresponding to the survey in year s is obtained, for example, by assuming that $\underline{y}_s$ from Equation (13.26) is multivariate normal with expectation $B\underline{N}_s$ and variance given by the variance-covariance matrix of the abundance estimates. If the integrated approach is used, the full likelihood from Section 7.2.2, corresponding to the survey in year s, may be used.

Note that there may be more than one observation equation in any given year, corresponding to more than one survey. In this case, the likelihood for year s is defined for each survey, and their product provides the required form for the observation component of the likelihood.

13.4 Model fitting

The fitting of models of the above type was until recently problematic. Any direct attempt to maximize the likelihood is impractical except in the simplest of cases, and often the likelihood may not have a unique maximum even when useful inference can be drawn. Three methods currently exist for fitting state models of population dynamics: the Kalman filter, Markov chain Monte Carlo (MCMC), and sequential importance sampling.

The Kalman filter (Kalman, 1960; Newman, 1998) provides a maximum likelihood method for fitting state-space models, and so it is more consistent with the rest of this book than the other methods. It is also the least computer-intensive. Its disadvantages are that the state-space model is assumed to be linear and multivariate normal, whereas the processes modelled are often markedly non-linear and non-normal.

The two remaining methods offer great flexibility. They both formulate the problem in a Bayesian framework, and simulate realizations of the population. Inference is drawn in a straightforward manner from the empirical distribution of each parameter of interest, estimated from these realizations. Both methods require the user to specify a joint prior distribution for $\underline{N}_0$ (the initial population size and structure) and ϕ. For the case of integrated modelling (as distinct from conditional modelling), a joint prior distribution must also be specified for $\underline{\theta}$.

Sequential importance sampling (Trenkel et al., 2000) takes advantage of the temporal factorization of the likelihood. Population parameters are simulated from the joint prior distribution, and populations are simulated from these parameter sets. In each survey year s, the likelihood is evaluated, given the data, for each parameter set and corresponding population. The parameter sets are then resampled with probability proportional to the corresponding likelihood. These are approximately a sample from the posterior distribution using data up to year s, which then provide the prior sample for year $s + 1$. After stepping through all years, the final sample of parameter sets is the required empirical posterior distribution, from which inference is drawn as follows. Any parameter of interest, such as current population size, or predicted future size, is evaluated using each parameter set in turn. The median of the resulting values provides an estimate of the parameter, and a 95% probability interval is estimated from the 2.5 and 97.5 percentiles of the ordered distribution of values.

Sequential importance sampling in this form may require excessive numbers of parameter sets to be drawn from the joint prior to ensure that sufficient different sets are retained in resamples to provide reliable estimation of the posterior. This problem may be avoided by perturbing the resampled parameter sets at each step, using kernel smoothing (Trenkel et al., 2000). Thus at each step, new parameter sets are 'spawned' in the area of parameter space supported by the data.

MCMC (Meyer and Millar, 1999a; Millar and Meyer, 2000a) provides an alternative fitting algorithm for state-space models (including non-linear, non-normal models). Although less straightforward conceptually, computer software (Meyer and Millar, 1999b) is available to allow simple implementation of the algorithm to state-space fisheries models (Millar and Meyer, 2000b). MCMC may also be used to obtain maximum likelihood estimates of the parameters (Newman, 2000).

13.5 Extensions

13.5.1 A multiple-survey method example

In our example, the survey method was line transect sampling, for which abundance may be estimated from a single survey. Multiple-survey methods may also be handled in the state-space framework. Suppose the survey method in our example had been a removal experiment, conducted in spring of each year, in which each animal has the same probability p of removal. In this scenario, the covered region is the whole survey region so the distribution matrix $\mathbf{D}$ does not apply, the state vector $\underline{N}_s$ does not contain states for animals outside the covered region and is therefore half the length it was in our previous example, and all other matrices are dimensioned accordingly. Suppose further that removed animals in year s can be tallied in four categories: y_{1ms} = males aged ≥ 1; y_{1fs} = females aged ≥ 1; y_{0ms} = juvenile males; and y_{0fs} = juvenile females. Then the state equation can be expressed as

$$E[\underline{N}_s] = \mathbf{XSRIZ}\underline{N}_{s-1} \tag{13.30}$$

with notation as before, apart from the addition of:

$$\mathbf{X} = \begin{bmatrix} 1-p & 0 & 0 & 0 & 0 & 0 \\ 0 & 1-p & 0 & 0 & 0 & 0 \\ 0 & 0 & 1-p & 0 & 0 & 0 \\ 0 & 0 & 0 & 1-p & 0 & 0 \\ 0 & 0 & 0 & 0 & 1-p & 0 \\ 0 & 0 & 0 & 0 & 0 & 1-p \\ p & 0 & 0 & 0 & 0 & 0 \\ 0 & p & 0 & 0 & 0 & 0 \\ 0 & 0 & p & 0 & 0 & 0 \\ 0 & 0 & 0 & p & 0 & 0 \\ 0 & 0 & 0 & 0 & p & 0 \\ 0 & 0 & 0 & 0 & 0 & p \end{bmatrix} \tag{13.31}$$

$$\mathbf{Z} = \begin{bmatrix} 1 & 0 & 0 & 0 & 0 & 0 & 0 & 0 & 0 & 0 & 0 & 0 \\ 0 & 1 & 0 & 0 & 0 & 0 & 0 & 0 & 0 & 0 & 0 & 0 \\ 0 & 0 & 1 & 0 & 0 & 0 & 0 & 0 & 0 & 0 & 0 & 0 \\ 0 & 0 & 0 & 1 & 0 & 0 & 0 & 0 & 0 & 0 & 0 & 0 \\ 0 & 0 & 0 & 0 & 1 & 0 & 0 & 0 & 0 & 0 & 0 & 0 \\ 0 & 0 & 0 & 0 & 0 & 1 & 0 & 0 & 0 & 0 & 0 & 0 \end{bmatrix} \tag{13.32}$$

Note that $\mathbf{X}$ splits each category of animal into two, corresponding to those animals retained and those removed. The vector $\underline{N}_s$ includes those removed animals, since they appear in the observation equation (below). Since they are not needed subsequently, $\mathbf{Z}$ removes them from the state vector, so that the vectors and matrices do not grow in size unnecessarily. In open-population mark-recapture by contrast, the state vector must be allowed to grow as the number of unique capture histories grows.

Returning to our example, the observation equation in year s now merely tallies number of animals removed by each observable category, and so is entirely deterministic:

$$
\underline{y}_s = \begin{bmatrix} y_{0ms} \\ y_{1ms} \\ y_{0fs} \\ y_{1fs} \end{bmatrix} = \begin{bmatrix} 1 & 0 & 0 & 0 & 0 & 0 \\ 0 & 1 & 1 & 0 & 0 & 0 \\ 0 & 0 & 0 & 1 & 0 & 0 \\ 0 & 0 & 0 & 0 & 1 & 1 \end{bmatrix} \begin{bmatrix} r_{0ms} \\ r_{1ms} \\ r_{2ms} \\ r_{0fs} \\ r_{1fs} \\ r_{2fs} \end{bmatrix} \tag{13.33}
$$

where r_{0ms} is the number of age 0 males removed in year s, and similarly for other terms; these 'removed' categories are the last six entries in the vector $\underline{N}_s$, which is of length 12.

Note that the removal experiment affects the states of the animals; the uncertainty associated with the experiment is thus in the state equation, not the observation equation. For example, r_{0ms} has a binomial distribution with parameters N_{0ms} and p, where N_{0ms} is the number of age 0 males in year s just before the removal experiment, and similarly for other terms.

13.5.2   Other extensions

Meta-populations (comprising distinct sub-populations with movement between them) can readily be accommodated. This can be done by adding a sub-population indicator to the state vector (this increases the length of the state vector by a multiple equal to the number of sub-populations) and introducing a movement matrix. (Equation (13.4) contains an example of a movement matrix.)

Survival, birth and (where relevant) movement rates can all be modelled as functions of covariates, as can probabilities of capture or detection.

If the surveys provide information on genotype of individuals, then categories of animal, corresponding to the different identifiable genotypes, may be defined. In cases where genetic theory specifies the probabilities that offspring belong to each genotype, given the parents, these probabilities may be incorporated in the state model, although a model for mating would also be required.

Species interactions, such as occur in predator-prey systems, or when two species compete for the same resource, are readily incorporated. In this case, there will be a section in $\underline{N}_s$ for each species. Models may then be specified for survival probabilities so that those for one species are a function of the abundance of other species.

13.6 Summary

The power and flexibility of state-space models, together with effective fitting algorithms, open up many options for wildlife management models. We have tried to indicate the promise of these methods, but much research is required before we can start to realize their full potential.

13.7 Exercises

Exercise 13.1   Specify the error vector $\underline{\varepsilon}_s$ explicitly for the model of Equation (13.4).

Exercise 13.2   Write out the likelihood of Equation (13.12) in full.

Exercise 13.3   Consider a series of S plot surveys of a closed population of N animals that are distributed uniformly and independently in a survey region of size A. The surveys involve repeatedly surveying the same plots, of total area a (and detecting all animals in these plots). Animals move about independently of each other, and in such a way that an animal's location on one survey occasion is independent of its location on previous occasions.

(a) Define appropriate states and write down the associated population vector, $\underline{N}_s$.

(b) For each state, write down the probabilities that an animal in the state on occasion $s - 1$ moves to each possible state on occasion s. (Hint: does the animal's state on one occasion depend on its state on previous occasions?)

(c) Hence specify the state model in matrix terms (similar to Equation (13.4)).

(d) Write down an appropriate observation model in matrix terms.

(e) Now consider the case in which each survey is a line transect survey. Given an estimate of the effective strip half-width for each survey (μ_s for $s = 1, \ldots, S$), write down an appropriate observation model in matrix terms (treating the μ_ss as known).

Exercise 13.4   Write down a state model in matrix form for the two-habitat movement model of Equation (13.4) when

(a) the probability that an animal is in habitat x on occasion s is constant and independent of where it was on occasion $(s-1)$ or on any previous occasion (this is an example of what we refer to as independent dynamics);

(b) the probability that an animal is in habitat x on occasion s is independent of where it was on occasion $(s-1)$ or on any previous occasion, but depends on occasion in the following way:

$$\pi_0(s) \;=\; \frac{1}{1 + e^{\phi_0 + \phi_s s}} \tag{13.34}$$

where $\pi_0(s)$ is the probability that an animal is in habitat $x = 0$ on occasion s and ϕ_0 and ϕ_s are parameters to be estimated (this is an example of what we refer to as population-level drift);

(c) there are known cumulative removals of $R_s(x)$ animals from habitat x on occasion s $(s = 1, ..., S; x = 0, 1)$.

Exercise 13.5   To see that $\underline{N}_s$ in Equation (13.26) is the vector of estimated numbers of juveniles, males and females in year s in the covered region, consider a line transect survey with covered area a and strip half-width w, in which n animals are detected.

(a) Given that an animal is in the covered region, and an estimate of effective strip half-width, write down an expression for the estimated probability that it is detected.

(b) Using the estimated probability from (a), write down a Horvitz–Thompson-like estimator for the number of animals in the covered region.

Exercise 13.6   In the example of Section 13.3, the only reason to model the male component of the population is that we have estimates of its size from the line transect surveys. Reformulate the state and observation equations of this example when:

(a) age 1 males are combined with males aged ≥ 2;

(b) each line transect survey yields a single estimate of female abundance, excluding age 0 females, and we wish to model the female component of the population only.

Exercise 13.7 For populations in which females have multiple births, what is the advantage of assuming the number of births per female follows a multinomial distribution rather than a Poisson distribution?

Exercise 13.8 Formulate the state and observation equations, and specify the error distributions, for each of the following examples. (Hint: pay careful attention to the time of year you wish $\underline{N_s}$ to correspond to in each case.)

(a) Newly-born animals have a summer survival rate ϕ_s, and the survival rate in the subsequent winter is ϕ_w. Older animals have an annual survival rate ϕ_a. Females first breed at age 2, and births V per adult female per year are distributed as multinomial, with $\Pr(V = 0) = 0.20$, $\Pr(V = 1) = 0.24$, $\Pr(V = 2) = 0.31$, $\Pr(V = 3) = 0.19$ and $\Pr(V = 4) = 0.06$, where these proportions have been estimated from a separate study. The probability that a newly-born animal is female is 0.5.

Births are assumed to occur in summer, followed by a single sample of a removal experiment in autumn. Animals are not sexed or aged on removal. Except for summer mortality of young, mortality is assumed to occur after the removal experiment, in winter and spring.

(b) Animals have a fixed survival rate ϕ_j in their first year, and an annual survival rate ϕ_a subsequently. Females first breed at age 1, and births per adult female per year are distributed as a binary variate with probability of success π. The probability that a newly-born animal is female is 0.5.

Births are assumed to occur in summer, followed by a count of the population in autumn. Juvenile animals can be separated from adults in the count, but animals cannot be sexed. The count can be assumed to be unbiased, and an estimate of the variance of the count is available. Mortality is assumed to occur primarily in winter and spring.

(c) First-year survival is assumed to be related to accumulated day degrees frost d as $\phi_j(d) = \exp(\alpha + \beta d)/[1 + \exp(\alpha + \beta d)]$ where $\beta < 0$. Subsequently, survival is assumed to be the following function of age a: $\phi(a) = \exp(\gamma + \delta \log a)/[1 + \exp(\gamma + \delta \log a)]$ up to age 3, after which the survival rate remains constant at the age 3 value. Births per female in year s are assumed to have a binary distribution with parameter $\pi(s) = \exp(\mu + \nu N_s)/[1 + \exp(\mu + \nu N_s)]$ where N_s is the total number of animals in the population just before the breeding season in year s, and $\nu < 0$. A newly-born animal is male with probability ψ. A line transect survey just after the breeding season yields estimates of the numbers of adult males, adult females, and young. A harvest takes place immediately after the survey, and numbers of animals removed are recorded by sex and by whether they are young or adult.

(d) As for part (b), animals have a fixed survival rate ϕ_j in their first year, and an annual survival rate ϕ_a subsequently. Females first breed at age 1, and births per adult female per year are distributed as a binary variate with probability of success π. The probability that a newly-born animal is female is 0.5.

Observations are assumed to be counts of females at breeding colonies. After winter mortality but before breeding, movement of young females only is assumed to occur between colonies. There are three colonies, and colony i has size $N_i(s)$ female animals and habitat suitability $H_i(s)$ in year s. The distance between colonies i and j is d_{ij}. The proportion of young (first-year) animals moving from colony i to colony j in year s is modelled as $\lambda_{ij} = \exp(\alpha + \beta N_i(s) + \gamma H_i(s) + \delta N_j(s) + \mu H_j(s) + \rho d_{ij})/[1 + \exp(\alpha + \beta N_i(s) + \gamma H_i(s) + \delta N_j(s) + \mu H_j(s) + \rho d_{ij})]$, where β and μ are positive and γ, δ and ρ are negative.

(e) For simplicity, populations are assumed to be exclusively female. Prey reproduce at rate λ and predators at rate μ in summer. Independent estimates of the sizes of the prey and predator populations are made just before breeding. Prey survival rate in year s is modelled as $\phi(s) = \exp(\alpha + \beta N(s) + \gamma P(s))/[1 + \exp(\alpha + \beta N(s) + \gamma P(s))]$ where $N(s)$ is the number of prey present at the start of the breeding season in year s, $P(s)$ is the corresponding number of predators, $\beta < 0$, and $\gamma < 0$. Predator survival rate in year s is modelled as $\psi(s) = \exp(\delta + \nu N(s) + \rho P(s))/[1 + \exp(\delta + \nu N(s) + \rho P(s))]$ where $\nu > 0$ and $\rho < 0$.

Part IV

Overview

14
Which method?

We have met a wide range of techniques for estimating the abundance of closed populations. When should we use which method? We give a few guidelines here. In any one field, one or two methods tend to dominate. For example, line transect sampling is the most commonly used method for estimating abundance of cetaceans and large terrestrial mammals; line and point transects for birds; mark-recapture and removal methods for small mammals; CPUE and other "harvest" methods for fisheries.

14.1   Plot sampling

Plot sampling is widely used in vegetation studies, where it is usually called "quadrat" sampling because the plots are quadrats. For species of plant in which individuals are readily identifiable, sample plots can readily be located through the survey region, and numbers of plants counted in each plot. The total count is then divided by the fraction of the survey area that was sampled, to provide an estimate of the number of plants in the survey region (Chapter 4). The method is cost-effective when the density of plants is very high and the survey area is small, so that the time to travel from one quadrat location to the next is small.

Apart from conventional quadrat sampling, plot sampling is used in various guises to estimate animal abundance. The plot may be an elongated rectangular strip, so that the observer can walk or travel from one end of the strip to the other, and count all animals within the strip. A number of strips are located within the study area, and estimation is exactly as described above. This method is useful for animals that occur at high density, so that animals detected outside the strip can be disregarded with little effect on precision. Animals should be easily detectable, so that it is possible to detect all (or almost all) animals within the sample strips. They should not move in response to the observer prior to detection, as this would bias the counts. When animals occur in large aggregations which spread over a large area, strip transect sampling can be effective. For example, in aerial surveys of walrus, when an aggregation is encountered, the covered strip is photographed, and numbers of animals in the strip counted later from the photographs. It would not be practical to attempt line transect methods on such populations.

For songbird surveys, the plot is often a circle, sometimes with a fixed radius, but sometimes with no specified radius. In the former case, abundance may be estimated as for standard quadrat sampling, although it is usually implausible that all birds within the circle will be detected. In practice, these "point count" methods are usually employed to estimate trends in relative abundance, rather than to provide estimates of absolute abundance. Even for this purpose, we would not recommend the method. Detectability typically varies by habitat, so that counts are not comparable across habitats. This in turn generates bias in estimates of trend over time if there is succession of habitat.

Another sample count method used widely for birds is the mapping census. Numbers of territories are recorded by site, where sites typically vary in size, and are subjectively selected. If the entire survey area is divided into sites and sites are randomly selected, absolute abundance may be estimated as for standard quadrat counts. As for point counts, the method is usually used for estimating trend in relative abundance. Again, we would not recommend the method, especially when sites are chosen subjectively, as trends on those sites may not be representative of the wider population. The method is also labour-intensive relative to transect methods, requiring several visits.

In the case of large terrestrial mammals that move away from the observer before they can be detected (e.g. deer or large cats in forest), plot sampling is often used to estimate dung density. This is then converted to an estimate of the number of animals that produced the dung. There are two approaches. Where dung density is sufficiently high, the size of the covered region may be small enough that it can be cleared of all dung prior to the survey, so that "clearance plot" methods can be used. The count takes place after a reasonable amount of dung has been deposited on the cleared plots but before any of that dung has decayed. Dung density can be converted to animal density by dividing by the product of the average number of droppings produced per day per animal and the number of days between clearance and the survey. If it is not feasible to clear plots, dung density is converted to animal density by dividing by the product of the average number of droppings produced per day per animal and the average number of days required for a dropping to decay. This is called the "standing crop" method. Generally, a larger area can be searched in a given time using strip transects of width 2–4 m rather than square quadrats (which are typically of side 5–10 m), as the observer can cover the strip in a single traverse.

14.2 Removal, catch-effort and change-in-ratio

The removal method (Chapter 5) has poor precision and the potential for large bias. It may have merit in a population that is to be controlled by constant-effort trapping over a short time period (so that the population can be assumed to be closed), as a crude means of estimating numbers of animals remaining in the population. However, even then the method performs poorly unless a large fraction of the population is removed. In addition, both heterogeneity and movement of animals into the survey region, as animals originally there are removed, can severely compromise estimates. If it is equally easy to mark captured animals and release them, then mark-recapture should be preferred to removal methods, as the additional information on recapture of marked animals allows abundance to be estimated with better precision, and improves ability to test the assumptions of the model. By releasing captured animals, we also avoid "sucking in" new animals from outside the survey region that tends to happen under the removal method.

Catch-effort methods are an extension of the removal method, in which the effort varies between occasions. The effort is assumed to be known, and it is assumed that we can model the relationship between catch and effort for a given population size. The method is particularly suited to harvested populations, although the problems in modelling the catch-effort relationship limit its value, especially if effort becomes more efficient (e.g. through use of improved technology) over time.

The change-in-ratio method, in common with the removal method, tends to have poor precision. It is nevertheless potentially a useful method for a harvested population in which an identifiable subset of animals is targeted in the harvest. This often occurs in trophy hunting, where hunters may be interested only in mature males. Culling of females may take place at a different time of year (if at all). In this circumstance, the ratio of males in the population can be estimated before and after the hunting season for males, or before and after the cull of females. There are essentially three requirements for the method to work well: that the numbers of animals culled are known for each category (e.g. males and females); that the targeted cull removes a high percentage of the targeted component of the population (perhaps more than 25%); and that the ratio of males in the population can he estimated with low bias both before and after the cull. For deer populations, it may be easier to classify animals as antlered or not, so that immature males are included with females. This reduces the problems of estimating the required ratio. If some non-targeted animals are taken in the targeted cull, the method remains valid, though some precision is lost.

Similar in character to the change-in-ratio method are clearance plot dung surveys, in which dung deposition per day is estimated before and after a large cull. If say 200 animals were culled, and the dung deposition per day within the survey region was estimated to have been reduced from 20,000 to 16,000, we would estimate that there were 1,000 animals in the population (of which 800 remain).

14.3   Mark-recapture

If it is practical to apply either method to a given population, mark-recapture (Chapters 6 and 11) has the following advantages over line transect sampling (Chapters 7 and 11): detailed information may be recorded on each captured animal; survival rates may be estimated; recaptures provide information on movement. In practice, most wildlife managers opt for line transect sampling if they have the choice of either. This is because mark-recapture field costs are substantially higher to achieve the same precision on abundance estimates, and typically, mark-recapture abundance estimates are more sensitive to failures of assumptions than line transect estimates. Indeed, some researchers maintain that mark-recapture cannot reliably estimate the abundance of biological populations, due to a high degree of heterogeneity which affects probability of capture and is difficult to model.

Mark-recapture may well provide the most reliable means of estimating the size of a closed population when the assumptions of line transect sampling are seriously violated. For populations comprising just a few tens of animals, such as coastal populations of bottle-nosed dolphins, it may be possible to identify a high proportion of animals, for example by catching and marking them, or by identifying natural marks on the animals. The number of known animals may itself be a useful estimate of minimum population size, and the effects of heterogeneity are typically rather small if 80–90% of the population are identifiable. Line transect sampling is not cost effective on such small populations, as it may be necessary to cover the area several times and perhaps detect every animal in the area several times before the population size can be quantified with acceptable precision.

Mark-recapture is also preferable to line transect sampling for populations of animals which may not be sufficiently detectable in a sightings survey, perhaps because they are small, are hidden amongst vegetation, move away from the observer before they can be detected or identified, or spend much time underground.

Note that the concept of "population" differs between single-survey methods such as line transect sampling and multiple-survey methods such as mark-recapture. In line transect sampling, we estimate the population that is in the survey region at the time of the survey. This "population" is well-defined, but a proportion of the population of interest may well be beyond the boundaries of the survey region. By contrast, the population being estimated in a mark-recapture study is rather more nebulous. It is the population of animals that might potentially be using the locations in which the traps are set at the time of the study. However, typically some of those animals will have very low probability of capture, perhaps because they seldom use the trapping locations, and as heterogeneity in probability of capture is notoriously difficult to model, we usually underestimate population size in this case. More pragmatically, we are estimating the size of the population whose members regularly use the survey area, but this population is not usually well-defined, and if transients pass through the population during the trapping period, upward bias in estimating this population size can be substantial.

14.4   Distance sampling

14.4.1   Line transect sampling

Line transect sampling (Section 7.2) may be viewed as an extension of strip transect sampling, in which not all animals within the strip need be detected. For the standard method, all animals on or very close to the line that bisects the strip lengthwise should be detected. The method is therefore useful for populations in which individuals are readily detectable if they are very close to the observer. If animals respond to the observer, this should only occur to any degree within a zone around the observer in which detection is nearly certain. Because there is no requirement that all animals off the line are detected, we can record animals in a wider strip than is possible with strip transect sampling. Thus the method is useful for sparsely distributed populations for which we should retain data on all or nearly all animals that we detect to achieve efficient estimation. The method is also particularly suited to surveys in which the survey region is large. A high proportion of time in the field is typically spent "on effort" (i.e. searching for animals from the line), so that the method makes efficient use of resources. Further, overall precision is largely determined by the number of animals detected, not by the size of the population. Precision is only affected by the size of the survey area in as much as resources are required to travel from one sample location to another.

14.4.2 Point transect sampling

Just as line transect sampling is an extension of strip transect sampling, point transect sampling (Section 7.3) is an extension of point counts. It is used almost exclusively for songbirds, although it has been applied for example to spotlight counts of hare and foxes and to estimating numbers of rare species of tree in tropical rain forest.

For songbirds, there are several advantages of point transect sampling over line transect sampling: in difficult terrain, it is easier to stand at a point and record birds rather than walk along a line; where access is difficult, it is easier to locate and get to a random point rather than navigate along a random line; for multispecies surveys in areas of high bird density, the observer can more easily get "swamped" when carrying out line transect surveys. There are also disadvantages: random movement of birds generates greater bias for point transects than for line transects; more time is spent "off-effort", travelling from one point to the next, reducing sample size; a larger sample size is needed for point transects than for line transects for a given precision, because the covered area close to the observer is smaller.

14.4.3 Cue counting

Cue counting (Section 7.4.1) was developed specifically for surveys of large whales. The survey design is very much as for line transect sampling, but all cues (usually blows) detected in a sector ahead of the observer are recorded, along with their distances from the observer. From these data, analyses closely related to point transect sampling yield an estimate of the number of cues per unit time per unit area, from which whale density is estimated, by dividing by the average number of cues per animal per unit time. The method does not require that all animals on the transect line are detected with certainty; rather, it assumes that cues that occur very close to the observer are certain to be detected. It is therefore better than standard (single-platform) line transect sampling when dive times of whales are sufficiently long that a proportion of animals on the trackline will remain undetected. Its main disadvantage is that the cue rate must be estimated, and doing this can be difficult: a closely-monitored animal may behave atypically, yielding a biased estimate of this quantity. Further bias arises because it is easier to monitor an animal that provides frequent cues, whereas an animal that dives for long periods may well be lost before its cue rate can be estimated.

14.4.4 Trapping webs

Trapping webs (Section 7.4.2) combine the concept of a trapping grid, for which the removal method may be applied to estimate abundance, and point transect sampling. Within a single web, traps are closely spaced at the centre, so that all animals close to the centre will be trapped. Trap separation increases with distance from the centre. The distance of the trap in which an animal is caught from the centre of the web is treated as the "detection distance" in a point transect analysis.

The method is potentially suitable for small mammals (especially rodents), reptiles and some insects (e.g. ground beetles). To estimate abundance within a wider survey region, it is necessary to establish a number of trapping webs, each of which may comprise well over 100 traps, so that the method is especially labour-intensive. It is also sensitive to animal movement, which leads to overestimation of abundance. Ideally, it should only be applied to animals that have strictly defined home ranges that are substantially smaller than the area covered by a single web, or that move only over small distances relative to the diameter of a single web.

14.4.5 Indirect distance sampling surveys

If a species is not readily surveyed, it is often worth considering whether they produce sign that is more readily surveyed using so-called indirect methods (Section 7.4.3; Buckland *et al.* (2001)). This approach can be effective for species that are too scarce to yield sufficient detections in say a distance sampling survey (e.g. apes and large cats), or for species that avoid detection by hiding or moving away from the observer (e.g. deer in closed habitats). The sign most commonly surveyed is dung (e.g. elephants, deer, large cats and foxes), although nests are typically surveyed in the case of apes. To estimate animal abundance, we must estimate the number of signs produced per animal per day. Except when sign is at very high density (when strip counts may prove more cost-effective), plots would not normally be cleared before they are surveyed, so that it is necessary also to estimate the mean decay rate of sign. Both rates can be problematic to estimate reliably, and decay rate especially may be highly variable, depending on conditions. These difficulties must be weighed against the difficulties in surveying the population using direct methods, to determine which approach can more easily yield abundance estimates with acceptable precision and low bias.

14.5   Point-to-nearest-object and nearest neighbour methods

In the point-to-nearest-object method (Chapter 8), a random point in the survey region is selected, and the distance to the nearest object (plant or animal) is measured. It is possible to estimate abundance from such distances. Similarly, in the nearest neighbour method, we select an object at random, and measure the distance to its nearest neighbour. These distances too allow estimation of abundance. Strong assumptions about the spatial distribution of objects are required in both cases. The methods have attracted some interest for populations of plants, but they are of limited practical value. Much time can be wasted in establishing which object is closest, and detections of other objects that turn out not to be the closest object are discarded. If distances to say the five closest objects are recorded, in an effort to improve efficiency, then it is difficult to establish which are the five closest objects. Further, in areas where the distribution of the population of interest is sparse, it may be virtually impossible to locate the nearest objects.

14.6   Migration counts

A special case of plot surveys is the migration count (Chapter 10), in which animals are counted as they migrate past a watch point. To date, the method has only been used for populations of large whale that migrate past coastal watch points. If detection is certain during watch periods, then we have a series of counts that sample time (the migration period) rather than space. These can be assumed to be a random sample of times from the migration period, from which abundance may be estimated in much the same way as for quadrat sampling. Alternatively, the rate of passage may be modelled as a function of date, allowing estimation of numbers of animals passing outside watch periods. (Both methods assume that passage rate at night is the same as during the day, unless an adjustment for night passage rate is made.)

14.7   Spatial modelling and integrated models

The distribution of animal populations is something that is often of interest to managers and biologists. Until recently, methods for estimating the spatial distribution of a population have largely been confined to estimating the abundances in a limited number of strata. The methods of Chapter 10 allow spatial distribution to be estimated at an arbitrarily small scale, allow the data to determine the shape of the estimated animal density surface, and allow density to be related to habitat type or other environmental variables. Spatial modelling methods have been used in relatively few applications and with a limited number of survey methods, but their use need not be limited. We expect to see increasing use of these methods, together with development of comparable methods for types of survey other than those covered in Chapter 10. There is potential for the development of useful spatial modelling with mark-recapture methods, for example. The ability of the methods to relate environmental variables to animal density is likely makes them particularly attractive in a world in which environmental change is increasingly important to wildlife population management.

To date, integrated models have not seen wide use, although there has been some use of methods involving integration of the various simple removal methods (Chapter 5), of spatial modelling methods for line transect surveys (Section 12.2), and of double-platform migration counts (Section 12.4).

Composite line transect and mark-recapture surveys are perhaps the one example of integrated modelling that has attracted appreciable research interest in recent years. When detection is not certain close to the line in line transect surveys, it has become standard to establish a second observation platform. This provides a two-sample mark-recapture experiment, where platform corresponds to sampling occasion (Section 12.3). We then record whether a detected animal is seen by the first platform alone, the second platform alone, or both platforms (a duplicate detection). The first "sampling occasion" need not correspond to an observation platform; it could instead correspond to a capture of animals, in which captured animals are released with radio tags. We then record during the line transect survey which of the animals with known locations are detected, again generating two-sample mark-recapture data. Another application of the method involves one visual detection occasion (e.g., a shipboard observer team) and one audio detection occasion (e.g., a towed hydrophone array).

Finally, we expect to see wide development of methods involving integration in open population contexts.

Appendix A
Notation and Glossary

A.1 Notation

We list below notation that is standard across all chapters and/or used in more than one chapter. Some notation is chapter-specific; you should look in the specific chapters for details of this notation.

$\pi(\ )$ probability density function (pdf) of animal-level variables. For example, if an animal's location is given by the coordinates (u, v), $\pi(u, v)$ is the pdf of animal location.

π_c coverage probability: the probability that a point in the survey region is in the covered region. In general this can be different for different points; for the point at (u, v) the coverage probability is written $\pi_c(u, v)$. Coverage probability is determined by the survey design. When it is the same at all points, we drop the brackets: π_c.

T indicates vector or matrix transpose. For example, $(x, y)^T = \begin{pmatrix} x \\ y \end{pmatrix}$.

$\hat{\ }$ a "hat" is used to denote an estimate or estimator. For example, $\hat{N}$ denotes an estimator or estimate of abundance.

θ parameter of the observation model. (In Chapter 7 it is also used to denote sighting angle in line transect surveys.)

$\underline{\theta}$ vector of parameters of the observation model.

ϕ parameter of the state model.

$\underline{\phi}$ vector of parameters of the state model.

A the surface area of the survey region.

a the surface area of the covered region.

AIC Akaike's information criterion (see Section 3.2.3).

B number of resamples in a bootstrap.

CI confidence interval.

CPUE catch per unit effort (see Section 5.3).

$E[\ ]$ expected value of the thing in square brackets.

$\mathcal{H}$ the "history" of all the relevant animal-level variables ($\underline{x}$s) of an animal over each survey occasion. $\mathcal{H}_{si}$ is animal is history up to the **start** of occasion s.

$\mathcal{H}_s$ a matrix whose rows are the histories of the individuals in the population ($\mathcal{H}_{s1}$ to $\mathcal{H}_{sN}$).

J number of survey design units in a sample; indexed $j = 1, \ldots, J$.

K number of animal types or strata; indexed $k = 1, \ldots, K$.

L likelihood function. Usually written with arguments (e.g., $L(N, p)$ for the likelihood function for N and p, or $L(N|n, p)$ for the likelihood function for N, given n and p). (L is also used to denote line length in line transect sampling in Chapter 7.)

ℓ log-likelihood function, $\ell = \ln(L)$. Usually written with arguments (e.g., $\ell(N|n, p)$ for the log-likelihood function for N, given n and p).

l_s survey effort value (if measured continuously) or type (if an indicator variable) on occasion s.

$\underline{l}_s$ vector of variables characterizing survey effort values and/or types on occasion s.

MLE maximum likelihood estimate or maximum likelihood estimator.

M_s number of marked animals in a mark-recapture survey at the **start** of occasion s.

m_s number of marked animals that are captured on occasion s in a mark-recapture survey.

N population size or abundance; the number of animals or groups in the population.

N_c the abundance of animals or groups in the covered region.

N_{cj} number of animals in survey unit j of the sample (covered region).

N_s population size or abundance on survey occasion s.

$N_s(\underline{x})$ the number of animals of type $\underline{x}$ in the population on survey occasion s.

$\underline{N}_s$ the vector of sub-population sizes when a population is composed of more than one type of animal. If males $(x = 0)$ and females $(x = 1)$ were distinguished, for example, $\underline{N}_s$ would be $(N_s(0), N_s(1))^T$.

n the number of animals detected or captured on a survey or over a number of surveys (depending on the context).

n_s the number of animals detected or captured on survey occasion s.

$n_s(\underline{x})$ the number of animals of type $\underline{x}$ that are detected (captured) on survey occasion s.

$\underline{n}_s$ the vector of numbers detected or captured of more than one type of animal. If males $(x = 0)$ and females $(x = 1)$ were distinguished, for example, $\underline{n}_s$ would be $(n_s(0), n_s(1))^T$.

p_{dj} probability that survey unit j is included in the sample when using a particular survey design.

p detection or capture probability. We assume that if animals are captured, all detected animals are captured.

$p(\ )$ detection or capture probability as a function of the variables appearing in the brackets.

q number of unknown parameters in a likelihood function.

pdf a probability density function (for continuous random variables), or probability mass function (for discrete random variables).

R_s cumulative removals by the **start** of survey occasion s.

$R_s(\underline{x})$ cumulative removals of animal type $\underline{x}$ by the **start** of survey occasion s.

$\underline{R}_s$ vector of cumulative removals of all animal types by the **start** of survey occasion s.

$\{n, R\}$ the sets of counts and removals by occasion (with only one animal type).

r_s number of animals removed on survey occasion s.

$r_s(\underline{x})$ number of animals of type $\underline{x}$ removed on survey occasion s.

$\underline{r}_s$ vector of numbers of animals of all types removed on survey occasion s.

S number of survey occasions; indexed $s = 1, \ldots, S$.

$se[\ ]$ the standard error of the thing in the square brackets, and equal to $\sqrt{Var[\ ]}$.

srs simple random sampling or simple random sample. Simple random sampling involves choosing survey units with equal probability and without replacement.

T total number of survey units in the survey region.

(u, v) Cartesian coordinates.

U_s number of unmarked animals in a mark-recapture survey at the **start** of occasion s $(U_1 = N)$.

u_s number of unmarked animals that are captured on occasion s in a mark-recapture survey.

$Var[\ ]$ the variance of the thing in the square brackets.

WiSP Wildlife Simulation software Package.

ω_{si} is the capture indicator for the ith animal on capture occasion s in a mark-recapture study: $\omega_{si} = 1$ if animal i is captured on occasion s, otherwise it is zero.

$\underline{\omega}_{si}$ is the capture history of the ith animal at the **start** of capture occasion s in a mark-recapture study: $\underline{\omega}_{si} = (\omega_{1i}, \ldots, \omega_{(s-1)i})$.

$\underline{x}$ vector of animal-level variables.

z group size.

A.2   Glossary

Coverage probability the probability that a point in the survey region is in the covered region. In general this can be different for different points; for the point at (u, v) the coverage probability is written $\pi_c(u, v)$. Coverage probability is determined by the survey design. When it is the same at all points, we drop the brackets: π_c.

Covered area the surface area of the covered region.

Covered region the region within which animals are at risk of being detected or captured – the region covered by the survey. The covered region is a subset of the survey region.

Estimate the value of an estimator on a particular occasion. The notation we use for estimates and estimators is the same. For example, $\hat{N} = 250$ is an estimate of N, and $\hat{N} = n/p$ is an estimator of N.

Estimator a function that converts data into an estimate. The notation we use for estimates and estimators is the same. For example, $\hat{N} = n/p$ is an estimator of N (it takes the datum, n, and returns an estimate of N), and $\hat{N} = 250$ is an estimate of N.

Heterogeneity differences in the detectability (or catchability) of individual animals. A population in which some animals are more detectable (catchable) than others is called a heterogeneous population. Note that a population can contain animals of very different types and be homogeneous in the sense we use the word, if all types of animal are equally detectable (catchable); heterogeneity refers to differences in detection (capture) probability. Note also that the word is used to refer to differences in detection (capture) probability due to some feature of the animals. For example, if the differences arise because of a feature of the survey (use of good observers sometimes and poor observers at other times, for example) we would not call the population heterogeneous.

Observation model statistical model describing how animals are detected or captured, given the searched region and their locations and characteristics.

Point estimate or estimator estimate or estimator of the value of a parameter.

Sampling distribution the pdf of a statistic (usually an estimator in this book) generated by sampling from some population. Figure 2.6 is an example.

State model statistical model describing the state of the population: the distribution and characteristics of animals in it.

Survey design the rules used to select the covered region; a particular covered region selected using these rules is referred to as a realization of the survey design.

Survey region the region within which we want to draw inferences about about animal abundance and/or distribution.

Survey area the surface area of the survey region.

Appendix B
Statistical formulation for observation models

This appendix contains a general statistical formulation for observation models pertinent to all the "simple" methods encountered in Part II. In it, we use the St Andrews example dataset for illustration. The vector of animal-level variables for this population is

$$
\underline{x} \;=\; \begin{pmatrix} \text{size} \\ \text{sex} \\ \text{exposure} \end{pmatrix} \tag{B.1}
$$

Similarly, $\underline{l}$ is in general a vector of survey-level variables. The variable $\underline{l}$ might be simply the number of observers used (in which case it is a scalar, not a vector), or if we wanted to estimate a separate detection probability for each observer, it could be the set of identities of the observers (as indicated by the observer numbers, for example).

Let $\underline{N}$ be the state vector giving the number of animals of each of K possible types in the population:

$$
\underline{N} \;=\; \begin{pmatrix} N(\underline{x}_1) \\ N(\underline{x}_2) \\ \vdots \\ N(\underline{x}_K) \end{pmatrix} \tag{B.2}
$$

With our example data, $K = 8$ sizes $\times$ 2 sexes $\times$ 2 exposure levels $= 32$, and

$$\underline{N} \;=\; \begin{pmatrix} N(\text{size}{=}1,\ \text{sex}{=}\text{male, exposed}) \\ N(\text{size}{=}1,\ \text{sex}{=}\text{male, unexposed}) \\ N(\text{size}{=}1,\ \text{sex}{=}\text{female, exposed}) \\ N(\text{size}{=}1,\ \text{sex}{=}\text{female, unexposed}) \\ \vdots \\ N(\text{size}{=}8,\ \text{sex}{=}\text{female, unexposed}) \end{pmatrix} \qquad \text{(B.3)}$$

Finally, define $\underline{n}$ similarly, to be the numbers of each type of animal detected or captured:

$$\underline{n} \;=\; \begin{pmatrix} n(\underline{x}_1) \\ n(\underline{x}_2) \\ \vdots \\ n(\underline{x}_K) \end{pmatrix} \qquad \text{(B.4)}$$

where $n(\underline{x})$ is the number of animals of type $\underline{x}$ which are detected.

Note that here we are considering only the case where $\underline{x}$ is discrete and its K values in the population are known.

We deal only with the case in which animals are detected/captured independently of one another (given their $\underline{x}$s and the associated $\underline{l}$s). In this case, we can express the observation model quite simply as the product of binomial models for each type of animal:

$$\Pr(\underline{n}\,|\,\underline{N},\underline{l}) \;=\; \prod_{k=1}^{K} \binom{N(\underline{x}_k)}{n(\underline{x}_k)} p(\underline{x}_k,\underline{l}_k)^{n(\underline{x}_k)}\,[1 - p(\underline{x}_k,\underline{l}_k)]^{N(\underline{x}_k)-n(\underline{x}_k)}$$

$$\text{(B.5)}$$

where $p(\ )$ is the detection function.

If $\underline{x}$ was continuous and every animal in the population has a unique $\underline{x}$, the above model becomes

$$\Pr(\underline{\omega}\,|\,\underline{N},\underline{l}) \;=\; \prod_{i=1}^{N} p(\underline{x}_i,\underline{l}_i)^{\omega_i}\,[1 - p(\underline{x}_i,\underline{l}_i)]^{1-\omega_i}$$

$$\text{(B.6)}$$

where $\underline{\omega} = (\omega_1,\ldots,\omega_N)^T$ is the vector of detection indicator variables for the population, with $\omega_i = 1$ if animal i was detected, and zero otherwise $(i = 1,\ldots,N)$.

B.1   Detection function

The observation model is incomplete without a specification for at least
the functional form of the detection function. Recall that the detection
function is the function which tells us the probability that an animal is
detected, given its characteristics, and the associated survey-level variables
used to search for it. It is rare that the detection function can validly be
treated as known, except when detection of animals within sampled units
is certain – as in quadrat surveys and strip transects. When it can't be
treated as known, the best we can do is to specify a sensible, adequately
flexible form, and use the survey data to estimate the unknown parameters
of the function.

A wide variety of functional forms are used, and the most appropriate
form depends on the context and the type of survey. For simplicity, we
often use the logistic form in this book. That is, we specify that

$$p(\underline{x}, \underline{l}) \quad = \quad \frac{e^{\theta_0 + \underline{x}'\underline{\theta}_x + \underline{l}'\underline{\theta}_l}}{1 + e^{\theta_0 + \underline{x}'\underline{\theta}_x + \underline{l}'\underline{\theta}_l}} \tag{B.7}$$

where

$$\underline{\theta} \quad = \quad \begin{pmatrix} \theta_0 \\ \underline{\theta}_x \\ \underline{\theta}_l \end{pmatrix} \tag{B.8}$$

is the vector of parameters of the detection function; θ_0 is an intercept
parameter, $\underline{\theta}_x$ is the vector of parameters associated with the animal-level
variables $\underline{x}$, and $\underline{\theta}_l$ is the vector of parameters associated with the survey-
level variables $\underline{l}$.

To take a relatively simple example, if $\underline{x} = z =$ size was the only relevant
animal-level variable, and $\underline{l} = l =$ number of observers the only relevant
survey-level variable, then $\underline{\theta}_x$ is the scalar θ_z, $\underline{\theta}_l$ is the scalar θ_l, and

$$p(\underline{x}, \underline{l}) \quad = \quad \frac{e^{\theta_0 + z\theta_z + l\theta_l}}{1 + e^{\theta_0 + z\theta_z + l\theta_l}} \tag{B.9}$$

might be an appropriate form for the detection function. Figure B.1 shows
an example of such a detection function.

Note that we use the term "detection function" to refer to the function
giving the probability of detecting the animal, no matter what the form of
the detection – sight, remote sensing, capture by trapping, etc.

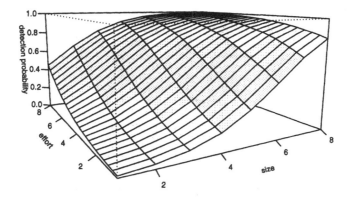

Figure B.1. An example of a two-dimensional logistic detection function.

B.2 Multiple surveys

Apart from plot sampling, distance sampling methods, and nearest neigh-bour methods, more than one survey is required in order to estimate N. We assume here that only one realization of the state model is involved; once assigned, the animal characteristics ($\underline{x}$) do not change over the course of the surveys. In this case, the single-survey observation model is easily extended to the case where there is more than one survey.

Removal methods involve the removal of a known number of animals from the population between the start of one survey occasion and the start of a subsequent one. Aside from any such known removals, the observation vector on survey occasion s, which we denote $\underline{n}_s$, is independent of obser-vation vectors on other occasions, so that the likelihood for a total of S surveys is just the product of the individual survey likelihoods.

To this end, we define a cumulative removals vector for occasion s:

$$\underline{R}_s = \begin{pmatrix} R_s(\underline{x}_1) \\ \vdots \\ R_s(\underline{x}_K) \end{pmatrix} \tag{B.10}$$

where $R_s(\underline{x}_k)$ is the total number of animals with animal-level variables $\underline{x}_k$ which have been removed by the start of survey occasion s (for $k = 1, \ldots, K$).

The number of animals with animal-level variables $\underline{x}_k$ which are still present in the population at the start of survey occasion s is then $N_s(\underline{x}_k)$, and the state vector on this occasion can be written as

$$\underline{N}_s \quad = \quad \underline{N} \quad - \quad \underline{R}_s$$

or

$$\begin{pmatrix} N_s(\underline{x}_1) \\ \vdots \\ N_s(\underline{x}_K) \end{pmatrix} = \begin{pmatrix} N(\underline{x}_1) \\ \vdots \\ N(\underline{x}_K) \end{pmatrix} - \begin{pmatrix} R_s(\underline{x}_1) \\ \vdots \\ R_s(\underline{x}_K) \end{pmatrix}$$

$$(B.11)$$

where $N(\underline{x}_k) = N_1(\underline{x}_k)$ is the initial number of animals with animal-level variables $\underline{x}_k$ in the population (for $k = 1, \ldots, K$).

With this notation we can write the likelihood for a series of S surveys as

$$\Pr(\boldsymbol{n} \mid \underline{N}, \boldsymbol{R}, \boldsymbol{l}) \quad = \quad \prod_{s=1}^{S} \Pr(\underline{n}_s \mid \underline{N}_s, \underline{l}_s)$$

$$= \quad \prod_{s=1}^{S} \prod_{k=1}^{K} \binom{N_s(\underline{x}_k)}{n_s(\underline{x}_k)} p_s(\underline{x}_k, \underline{l}_s)^{n_s(\underline{x}_k)}$$
$$\times \, [1 - p_s(\underline{x}_k, \underline{l}_s)]^{N_s(\underline{x}_k) - n_s(\underline{x}_k)}$$

$$(B.12)$$

Here

- $\boldsymbol{n} = (\underline{n}_1, \ldots, \underline{n}_S)$ is the matrix whose columns are the observation vectors for each survey,

- $\boldsymbol{R} = (\underline{R}_1, \ldots, \underline{R}_S)$ is the matrix whose columns are the cumulative removal vectors for each survey,

- $\boldsymbol{l} = (\underline{l}_1, \ldots, \underline{l}_S)$ is the matrix whose columns are the vectors defining the survey-level variables for each survey.

Appendix C
The asymptotic variance of MLEs

In this appendix we outline some theoretical results that underlie estimation of the asymptotic variance of MLEs and functions of MLEs. A fuller treatment can be found in Cox and Hinkley (1974) or Silvey (1975).

C.1   Estimating the variance of an MLE

Let $l(\underline{\theta})$ be the log-likelihood function of t unknown parameters $\underline{\theta}$ and let $l_i(\underline{\theta})$ be the contribution to the log-likelihood for the ith observation ($i = 1, 2, \ldots, n$). We assume that the observations are independently distributed, from which it follows that

$$l(\underline{\theta}) \;=\; \sum_{i=1}^{n} l_i(\underline{\theta}) \tag{C.1}$$

The maximum likelihood estimator of $\underline{\theta}$ is the vector $\hat{\underline{\theta}}$ that maximizes $l(\underline{\theta})$, i.e. $l(\hat{\underline{\theta}}) \geq l(\underline{\theta})$ for all permissible vectors $\underline{\theta}$.

Usually, it is not possible to give exact expressions for the variances of the MLEs $\hat{\theta}_1, \hat{\theta}_2, \ldots, \hat{\theta}_t$. However, there is a general method of getting asymptotic[1] approximations for the variances. It is based on the "information matrix". There are two types of information matrix: the "observed

---

[1] That is, the approximation improves as sample size n increases, and becomes "exact" for infinite n.

information matrix" (which we denote $I(\underline{\theta})$), and the "Fisher information matrix" (which we denote $i(\underline{\theta})$). For very large sample sizes, these lead to equivalent results but in many cases it is difficult to assess which one is the more accurate for smaller sample sizes.

The (j, k)th entry of the observed information matrix, $I(\underline{\theta})$, is given by

$$I_{j,k}(\underline{\theta}) \;=\; -\sum_{i=1}^{n} \frac{d^2 l_i(\underline{\theta})}{d\theta_j d\theta_k} \qquad (j, k = 1, 2, \ldots, t) \qquad \text{(C.2)}$$

The Fisher information matrix is the expected value of $I(\underline{\theta})$, i.e.

$$i(\underline{\theta}) \;=\; E\left[I(\underline{\theta}) \right] \qquad \text{(C.3)}$$

An alternative and often simpler method of getting $i(\underline{\theta})$ is to use the result that under certain regularity conditions

$$i_{j,k}(\underline{\theta}) \;=\; E\left[\frac{dl_i(\underline{\theta})}{d\theta_j} \frac{dl_i(\underline{\theta})}{d\theta_k} \right] \qquad \text{(C.4)}$$

where $i_{j,k}(\underline{\theta})$ is the (j, k)th entry of $i(\underline{\theta})$.

A standard result in asymptotic theory states that (under certain regularity conditions) the MLE $\hat{\underline{\theta}}$ is asymptotically normally distributed with expected value $\underline{\theta}_0$ and variance-covariance matrix $i(\underline{\theta}_0)^{-1}$, where $\underline{\theta}_0$ is the vector of (unknown) true values of the parameters. So the MLE is asymptotically unbiased and the variance of each $\hat{\theta}_j$ can be obtained from the information matrix.

If, for example, there are two unknown parameters $(t = 2)$,

$$i(\underline{\theta}_0)^{-1} \;=\; \left(\begin{array}{cc} Var[\hat{\theta}_1] & Cov[\hat{\theta}_1 \hat{\theta}_2] \\ Cov[\hat{\theta}_1 \hat{\theta}_2] & Var[\hat{\theta}_2] \end{array} \right) \qquad \text{(C.5)}$$

In general, $Var[\hat{\theta}_j]$ is the jth diagonal element of the matrix $i(\underline{\theta}_0)^{-1}$.

You can use either the jth diagonal element of $i(\hat{\theta})^{-1}$ or the jth diagonal element of $I(\hat{\theta})^{-1}$ to estimate $Var[\hat{\theta}_j]$.

In cases where the MLE $\hat{\underline{\theta}}$ can't be expressed analytically, one has to use numerical methods to compute the estimates. Similarly, it is often not possible to obtain expressions for $i(\hat{\theta})$. However, some software packages used to calculate $\hat{\underline{\theta}}$ numerically, supply the estimated information per observation, which is equal to $\frac{1}{n} i(\hat{\theta})$.

C.2   Estimating the variance of a function of an MLE

In the simplest case the abundance, N, is one of the elements of $\underline{\theta}$, but often the MLE of abundance is a function of $\hat{\underline{\theta}}$ and we can write it as $\hat{N} = g(\hat{\underline{\theta}})$, where $g()$ is the appropriate function. A standard result in asymptotic theory states that (under certain regularity conditions) $g(\hat{\underline{\theta}})$ is asymptotically normal with mean $g(\underline{\theta}_0)$ and variance

$$\left[\frac{dg(\underline{\theta})}{d\underline{\theta}}\bigg|_{\underline{\theta}=\underline{\theta}_0} \right]^T i(\underline{\theta})_0^{-1} \left[\frac{dg(\underline{\theta})}{d\underline{\theta}}\bigg|_{\underline{\theta}=\underline{\theta}_0} \right] \tag{C.6}$$

We estimate this either by

$$\widehat{Var}[\hat{N}] = \left[\frac{dg(\underline{\theta})}{d\underline{\theta}}\bigg|_{\underline{\theta}=\hat{\underline{\theta}}} \right]^T i(\hat{\underline{\theta}})^{-1} \left[\frac{dg(\underline{\theta})}{d\underline{\theta}}\bigg|_{\underline{\theta}=\hat{\underline{\theta}}} \right] \tag{C.7}$$

or by this equation with $I(\hat{\underline{\theta}})^{-1}$ in place of $i(\hat{\underline{\theta}})^{-1}$.

C.3   A one-parameter example

Suppose m_2 marked animals are captured when sampling from a population containing n_1 marked animals. Suppose also that m_2 is binomially distributed with parameters n_1 (which is known) and p (which is unknown). In this case we have only one unknown parameter: $\theta = p$. The likelihood is

$$L(p) = \binom{n_1}{m_2} p^{m_2} (1-p)^{n_1-m_2} \tag{C.8}$$

from which we get the log-likelihood

$$l(p) = \ln(n_1!) - \ln((n_1-m_2)!) - \ln(m_2!) \\ + m_2 \ln(p) + (n_1-m_2)\ln(1-p) \tag{C.9}$$

Differentiating with respect to p and setting the derivative to zero, we get

$$\frac{dl(p)}{dp} = 0 \Rightarrow \frac{m_2}{p} = \frac{n_1-m_2}{1-p} \Rightarrow \hat{p} = \frac{m_2}{n_1} \tag{C.10}$$

C.3.1 Fisher information version 1

$$
\begin{aligned}
i(\underline{\theta}) &= E\left[\left(\frac{dl(p)}{dp}\right)^2\right] = \frac{E\left[(m_2 - n_1 p)^2\right]}{p^2(1-p)^2} \\
&= \frac{n_1 p(1-p)}{p^2(1-p)^2} = \frac{n_1}{p(1-p)}
\end{aligned}
$$

(C.11)

(C.12)

Hence the asymptotic variance of $\hat{p}$ is

$$
Var\left[\hat{p}\right] = \frac{p(1-p)}{n_1}
$$

(C.13)

which can be estimated using

$$
\widehat{Var}\left[\hat{p}\right] = \frac{m_2(n_1 - m_2)}{n_1^3}
$$

(C.14)

C.3.2 Fisher information version 2

$$
\begin{aligned}
i(\underline{\theta}) &= -E\left[\frac{d^2 l(p)}{dp^2}\right] = -E\left[\frac{n_1 p(1-p) + (m_2 - n_1 p)(1-2p)}{p^2(1-p)^2}\right] \\
&= \frac{n_1 p(1-p)}{p^2(1-p)^2} = \frac{n_1}{p(1-p)}
\end{aligned}
$$

(C.15)

as above.

C.3.3 Observed information

$$
\begin{aligned}
I(\underline{\theta}) &= -\sum_{i=1}^{n} \frac{d^2 l_i(p)}{d^2 p} = \frac{m_2}{p^2} - \frac{(n_1 - m_2)}{(1-p)^2} \\
&= \frac{(2m_2 - n_1)^2 p^2 - 2m_2 p + m_2}{p^2(1-p)^2}
\end{aligned}
$$

(C.16)

Replacing p by the MLE $\hat{p} = \frac{m_2}{n_1}$ gives

$$
I(\hat{p}) = \frac{n_1^3}{m_2(n_1 - m_2)}
$$

(C.17)

Hence

$$
\widehat{Var}\left[\hat{p}\right] = \frac{m_2(n_1 - m_2)}{n_1^3}
$$

(C.18)

which is identical to Equation (C.14).

C.3.4   Estimating the variance of $\hat{N}$

The Petersen mark-recapture estimator of abundance is the function $\hat{N} = g(\hat{p}) = \frac{n_2}{\hat{p}}$. Now

$$\frac{dg(p)}{dp}\bigg|_{p=m_2/n_1} = \frac{-n_2}{p^2}\bigg|_{p=m_2/n_1} = \frac{-n_2 n_1^2}{m_2^2} \tag{C.19}$$

so, using Equation (C.7), we estimate the variance of $\hat{N}$ by

$$\widehat{Var}\left[\hat{N}\right] = \left[\frac{dg(p)}{dp}\bigg|_{p=m_2/n_1}\right]^T \widehat{Var}\left[\hat{p}\right]\left[\frac{dg(p)}{dp}\bigg|_{p=m_2/n_1}\right]$$

$$= \left[\frac{-n_2 n_1^2}{m_2^2}\right]\frac{m_2(n_1 - m_2)}{n_1^3}\left[\frac{-n_2 n_1^2}{m_2^2}\right]$$

$$= \frac{n_1 n_2^2(n_1 - m_2)}{m_2^3} \tag{C.20}$$

Appendix D
State models for mark-recapture and removal methods

In this appendix, we derive state models for the animals with unknown xs in mark-recapture and removal studies. We talk in terms of a marked population; in removal studies a marked animal is a removed animal. We consider three kinds of state model: a static population model, a model for populations with independent dynamics, and one for populations with Markov dynamics.

D.1 Static population

With a static population, the xs in the population can be thought of as being determined before the first survey and not changing. Because the most detectable animals are removed first, the xs of the unmarked animals tend to contain a higher and higher proportion of animals with xs that make them hard to detect. The distribution of the xs of unmarked animals is given by the conditional density of x, given that the animal is unmarked by the start of survey s (i.e. that $\underline{\omega}_{si} = \underline{0}_s$). It is

$$\pi(x|\underline{0}_s) = \frac{P(\underline{0}_s \,|\, x)\pi(x)}{E\left[P(\underline{0}_s \,|\, x)\right]} \tag{D.1}$$

where $P(\underline{0}_s \,|\, x)$ is the probability that an animal is uncaptured by the start of capture occasion s (given x and the survey-level variables applied on each survey), and $E[P(\underline{0}_s \,|\, x)]$ is the expected value of $P(\underline{0}_s \,|\, x)$ with respect to x:

$$P(\underline{0}_s \,|\, \underline{x}) \;=\; \prod_{s^*=1}^{s-1} (1 \;-\; p(\underline{x}, \underline{0}_{s^*}, \underline{l}_{s^*})) \tag{D.2}$$

$$E[P(\underline{0}_s \,|\, \underline{x})] \;=\; \int P(\underline{0}_s \,|\, \underline{x}) \pi(\underline{x}) \, d\underline{x} \tag{D.3}$$

$$=\; \int \prod_{s^*=1}^{s-1} [1 \;-\; p(\underline{x}, \underline{0}_{s^*}, \underline{l}_{s^*})] \pi(\underline{x}) \, d\underline{x}$$

Now the probability that an uncaptured animal that has animal-level variables $\underline{x}$ is missed on occasion s can be written as

$$P(\underline{0}_{(s+1)} | \underline{0}_s, \underline{x}) \;=\; [1 - p(\underline{x}, \underline{0}_s, \underline{l}_s)] \tag{D.4}$$

From this and Equation D.1, it follows that the probability that an uncaptured animal (with unknown $\underline{x}$) is missed on occasion s is

$$E\left[P(\underline{0}_{(s+1)} | \underline{0}_s, \underline{x})\right] \;=\; \int [1 - p(\underline{x}, \underline{0}_s, \underline{l}_s)] \, \pi(\underline{x} | \underline{0}_s) \, d\underline{x}$$

$$=\; \frac{E\left[P(\underline{0}_{(s+1)} | \underline{x})\right]}{E\left[P(\underline{0}_s | \underline{x})\right]} \tag{D.5}$$

D.2 Independent dynamics

This section deals with models in which $\underline{x}_i$ is determined on each occasion independently of its values on other occasions. In the woodland and grassland example, this would be the case if the probability that an animal is in woodland on capture occasion s is ϕ on every occasion, no matter where it was on previous occasions. Because $\underline{x}_{si}$ is independent of all previous $\underline{x}_{s^*i}$ ($s^* < s$), the conditional probability density of $\underline{x}_{si}$, given capture history $\underline{\omega}_{si}$, is just $\pi(\underline{x}_{si})$. That is,

$$\pi(\underline{x}_{si} | \underline{\omega}_{si}) \;=\; \pi(\underline{x}_{si}) \tag{D.6}$$

The probability that animal i with capture history $\underline{\omega}_{si}$ is missed on occasion s (i.e. has $\omega_{si} = 0$) and has animal-level variables $\underline{x}_{si}$ can therefore be written as

$$P(0, \underline{x}_{si} | \underline{\omega}_{si}) \;=\; [1 - p(\underline{x}_{si}, \underline{\omega}_{si}, \underline{l}_s)] \pi(\underline{x}_{si})$$

$$\tag{D.7}$$

(For brevity we write $P(\omega_{si} = 0, \underline{x}_{si} | \underline{\omega}_{si})$ as $P(0, \underline{x}_{si} | \underline{\omega}_s)$.) From this it follows that the probability that an animal with capture history $\underline{\omega}_s$ (and unknown $\underline{x}_s$) is missed on occasion s is

$$
\begin{aligned}
E\left[P(0 | \underline{\omega}_s)\right] &= E\left[P(0, \underline{x}_s | \underline{\omega}_s)\right] \\
&= \int \left[1 - p(\underline{x}_s, \underline{\omega}_s, \underline{l}_s)\right] \pi(\underline{x}_s)\, d\underline{x}_s
\end{aligned}
\tag{D.8}
$$

(For brevity we write $E\left[P(\omega_s = 0, \underline{x}_s | \underline{\omega}_s)\right]$ as $E\left[P(0 | \underline{\omega}_s)\right]$.)

If animals are individually identifiable, then at the start of capture occasion s we know the number of animals with each possible capture history, except for those with capture history $\underline{0}_s$. Let $N_s(\underline{\omega}_s)$ be the number of animals with capture history $\underline{\omega}_s$ (so $U_s = N_s(\underline{0}_s)$). If $n_s(\underline{\omega}_s)$ of these are captured on occasion s, the contribution to the full likelihood from animals with capture history $\underline{\omega}_s$ is

$$
\begin{aligned}
L_{s\underline{\omega}_s} &= \binom{N_s(\underline{\omega}_s)}{n_s(\underline{\omega}_s)} E[P(0 | \underline{\omega}_s)]^{N_s(\underline{\omega}_s) - n_s(\underline{\omega}_s)} \\
&\quad \times \prod_i p(\underline{x}_{si}, \underline{\omega}_{si}, \underline{l}_s) \pi(\underline{x}_{si})
\end{aligned}
\tag{D.9}
$$

where the product is over all animals with capture history $\underline{\omega}_s$ that are captured on occasion s.

The full likelihood for capture occasion s is the product of these likelihood components over all possible capture histories:

$$
L_s = \prod_{\underline{\omega}_s} L_{s\underline{\omega}_s}
\tag{D.10}
$$

and the full likelihood is the product over all capture occasions:

$$
L = \prod_{s=1}^{S} L_s
\tag{D.11}
$$

D.3   Markov dynamics

Recall that $\mathcal{H}_{si}$ is the history of the animal-level variables of the ith animal up to the start of survey occasion s (Chapter 9). If we let $l_{si} =$

$(\underline{l}_{1i}, \ldots, \underline{l}_{(s-1)i})$ represent the (known) survey-level variables that applied to animal i on the first $s-1$ survey occasions, then the probability of getting the capture history $\underline{\omega}_{si}$, given the animal's history $\mathcal{H}_{si}$ and l_{si} can be written as

$$
\begin{aligned}
P(\underline{\omega}_{si}|\mathcal{H}_{si}) &= P(\underline{\omega}_{si}|\mathcal{H}_{si}, l_{si}) \qquad\qquad\text{(D.12)} \\
&= \prod_{s^*=1}^{s-1} p(\underline{x}_{si})^{\omega_{si}} [1 - p(\underline{x}_{si})]^{1-\omega_{si}}
\end{aligned}
$$

(For brevity, we write $P(\underline{\omega}_{si}|\mathcal{H}_{si}, l_{si})$ as $P(\underline{\omega}_{si}|\mathcal{H}_{si})$ and $p(\underline{x}_{si}, \underline{\omega}_{si}, l_{si})$ as $p(\underline{x}_{si})$.)

This is the observation equation for occasions $1, \ldots, s$. But unless animal i is detected on every occasion prior to s, we do not observe $\mathcal{H}_{si}$ and we can't actually evaluate the observation equation. In the case of the U_s animals that are uncaptured by the start of occasion s, we do not observe $\mathcal{H}_{si}$ at all, while in the case of animals that are captured on some but not all occasions prior to s, we observe some but not all of $\mathcal{H}_{si}$. Let $\mathcal{H}_{si}^o$ be the observed component, and $\mathcal{H}_{si}^u$ be the unobserved component.

We can now write the conditional probability density function of $\underline{x}_{si}$, given the observed history, $\mathcal{H}_{si}^o$, as follows. (Note that if we know the observed history $\mathcal{H}_{si}^o$, we also know the capture history $\underline{\omega}_{si}$, so it is sufficient to condition on $\mathcal{H}_{si}^o$.)

$$
\begin{aligned}
\pi(\underline{x}_{si}|\mathcal{H}_{si}^o) &= \frac{\left[\int P(\underline{\omega}_{si}|\mathcal{H}_{si})\pi(\mathcal{H}_{si})\pi(\underline{x}_{si}|\underline{x}_{(s-1)i})d\mathcal{H}_{si}^u\right]}{\int \left[\int P(\underline{\omega}_{si}|\mathcal{H}_{si})\pi(\mathcal{H}_{si})\pi(\underline{x}_{si}|\underline{x}_{(s-1)i})d\mathcal{H}_{si}^u\right] d\underline{x}_{si}} \\
&= \frac{\left[\int P(\underline{\omega}_{si}|\mathcal{H}_{si})\pi(\mathcal{H}_{si})\pi(\underline{x}_{si}|\underline{x}_{(s-1)i})d\mathcal{H}_{si}^u\right]}{E\left[P(\underline{\omega}_{si}|\mathcal{H}_{si}^o)\right]}
\end{aligned}
$$

$$\text{(D.13)}$$

where, if $\underline{x}$ is continuous, $\int \ldots d\mathcal{H}_{si}^u$ is the integral over the unobserved $\underline{x}_{s^*i}$ for $s^* < s$, while if $\underline{x}$ is discrete it is the sum over the unobserved $\underline{x}_{s^*i}$ for $s^* < s$.

The probability that an animal with observed history $\mathcal{H}_{si}^o$ is missed on occasion s and has animal-level variables $\underline{x}_{si}$ on this occasion is

$$
P(0, \underline{x}_{si}|\mathcal{H}_{si}^o) = [1 - p(\underline{x}_{si}, \underline{\omega}_{si}, \underline{l}_s)]\pi(\underline{x}_{si}|\mathcal{H}_{si}^o)
$$

$$\text{(D.14)}$$

It follows that the probability that an animal with observed history $\mathcal{H}_{si}^o$ and unknown $\underline{x}_s$ is missed on occasion s is

$$E\left[P(0|\mathcal{H}_{si}^{o})\right] \;=\; \int P(0, \underline{x}_s | \mathcal{H}_{si}^{o}) \, d\underline{x}_s$$

$$(\text{D}.15)$$

In the case of animals that are uncaptured by the start of occasion s, $\mathcal{H}_s^o$ is no more than the capture history $\underline{0}_s$ and $E[P(0|\mathcal{H}_{si}^o)] = E[P(\underline{0}_{(s+1)}|\underline{0}_s)]$. The component of the likelihood on occasion s for these animals is very similar to the likelihood Equation (D.9) of the previous section:

$$L_{s0} \;=\; \binom{N_s(\underline{0}_s)}{n_s(\underline{0}_s)} E[P(\underline{0}_{(s+1)}|\underline{0}_s)]^{N_s(\underline{0}_s)-n_s(\underline{0}_s)}$$

$$\times \prod_i p(\underline{x}_{si}, \underline{0}_{si}, \underline{l}_s) \pi(\underline{x}_{si}|\underline{0}_s)$$

$$(\text{D}.16)$$

where the product is over all animals that are captured for the first time on occasion s.

References

Agresti, A. 1994. Simple capture-recapture models permitting unequal catchability and variable sampling effort. *Biometrics*, **50**, 494–500.

Akaike, H. 1973. Information theory and an extension of the maximum likelihood principle. *Pages 267–281 of:* B.N., Petran, and Csàaki, F. (eds), *International symposium on information theory*. Akadèemiai Kiadi.

Alho, J.M. 1990. Logistic regression in capture-recapture models. *Biometrics*, **46**, 623–635.

Alpizar-Jara, R., and Pollock, K.H. 1999. Combining line transect and capture-recapture for mark-resightings studies. *Pages 99–114 of:* Garner, G.W., Amstrup, S.C., Laake, J.L., Manly, B.F.J., McDonald, L.L., and Robertson, D.G. (eds), *Marine mammal survey and assessment methods*. Rotterdam: Balkema.

Ashbridge, J., and Goudie, I.B.J. 2000. Coverage-adjusted estimators for mark-recapture in heterogeneous populations. *Communications in Statistics: Simulation and Computation*, **29**, 1215–1237.

Augustin, N.H., Mugglestone, M.A., and Buckland, S.T. 1996. An autologistic model for the spatial distribution of wildlife. *Journal of Applied Ecology*, **33**, 339–347.

Augustin, N.H., Mugglestone, M.A., and Buckland, S.T. 1998a. The role of simulation in modeling spatially correlated data. *Environmetrics*, **9**, 175–196.

Augustin, N.H., Borchers, D.L., Clarke, E.D., Buckland, S.T., and Walsh, M. 1998b. Spatiotemporal modelling for the annual egg production method of stock assessment using generalized additive models. *Canadian Journal of Fisheries and Aquatic Sciences*, **55**, 2608–2621.

Borchers, D.L. 1996. *Line transect abundance estimation with uncertain detection on the trackline.* PhD Thesis, University of Cape Town, Cape Town.

Borchers, D.L., and Cameron, C. unpublished. *Analysis of the 1992/93 IWC minke whale sightings survey in Area III.* Paper SC/47/SH16 presented to the IWC Scientific Committee, 1995.

Borchers, D.L., Buckland, S.T., Priede, I.G., and Ahmadi, S. 1997. Improving the precision of the daily egg production method using generalized additive models. *Canadian Journal of Fisheries and Aquatic Sciences*, **54**, 2727–2742.

Borchers, D.L., Buckland, S.T., Goedhart, P.W., Clarke, E.D., and Hedley, S.L. 1998a. Horvitz-Thompson estimators for double-platform line transect surveys. *Biometrics*, **54**, 1221–1237.

Borchers, D.L., Zucchini, W., and Fewster, R. 1998b. Mark-recapture models for line transect surveys. *Biometrics*, **54**, 1207–1220.

Bowman, A.W., and Azzalini, A. 1997. *Applied smoothing techniques for data analysis.* Oxford: Clarendon Press.

Branch, T.A., and Butterworth, D.S. 2001. Southern Hemisphere minke whales: standardised abundance estimates from the 1978/79 to 1997/98 IDCR-SOWER surveys. *Journal of Cetacean Research and Management*, **3**(2), 143–174.

Bravington, M. 1994. *The effects of acidification on the population dynamics of brown trout in Norway.* PhD Thesis, Imperial College, London.

Bravington, M.V. unpublished. *Covariate models for continuous time sightings data.* Paper SC/52/IRMP14 presented to the International Whaling Commission Scientific Committee, 2000.

Breiwick, J.M., Rugh, D.J., Withrow, D.E., Dalheim, M.E., and Buckland, S.T. unpublished. *Preliminary population estimate of gray whales during the 1987/88 southward migration.* Paper SC/49/PS12 presented to the International Whaling Commission Scientific Committee, 1998.

Buckland, S.T., and Breiwick, J.M. in press. Estimated trends in abundance of California gray whales from shore counts, 1967/68 to 1987/88. *Journal of Cetacean Research and Management.*

Buckland, S.T., Breiwick, J.M., Cattanach, K.L., and Laake, J.L. 1993. Estimated population size of the California gray whale. *Marine Mammal Science*, **9**, 235–249.

Buckland, S.T., Burnham, K.P., and Augustin, N.H. 1997. Model selection: an integral part of inference. *Biometrics*, **53**, 603–618.

Buckland, S.T., Anderson, D.R., Burnham, K.P., Laake, J.L., Borchers, D.L., and Thomas, L.J. 2001. *Introduction to distance sampling.* Oxford: Oxford University Press.

Burnham, K.P. 1972. *Estimation of population size in multinomial capture-recapture studies when capture probabilities vary among animals.* PhD Thesis, Oregon State University, Oregon.

Burnham, K.P., and Overton, W.S. 1978. Estimation of the size of a closed population when capture probabilities vary among animals. *Biometrika*, **75**, 625–633.

Burnham, K.P., and Overton, W.S. 1979. Robust estimation of population size when capture probabilities vary among animals. *Ecology*, **60**, 927–936.

Burnham, K.P., Anderson, D.R., White, G.C., Brownie, C., and Pollock, K.H. 1987. *Design and analysis methods for fish survival experiments based on release-recapture.* American Fisheries Society Monographs, no. 5. American Fisheries Society.

Chao, A. 1987. Estimating the population size for capture-recapture data with unequal capture probabilities. *Biometrics*, **43**, 783–791.

Chao, A. 1989. Estimating population size for sparse data in capture-recapture experiments. *Biometrics*, **45**, 427–438.

Chao, A., Lee, S.M., and Jeng, S.L. 1992. Estimating population size for capture-recapture data when capture probabilities vary by time and individual animal. *Biometrics*, **48**, 201–216.

Chapman, D.G. 1951. Some properties of the hypergeometric distribution with applications to zoological censuses. *University of California Publications in Statistics*, **1**, 131–160.

Chapman, D.G., and Junge, C.O. 1956. The estimation of the size of a stratified animal population. *Annals of Mathematical Statistics*, **27**, 375–389.

Chapman, D.G., and Murphy, G.I. 1965. Estimates of mortality of the size of a stratified animal population. *Annals of Mathematical Statististics*, **27**, 375–389.

Chen, C.L., Pollock, K.H., and Hoenig, J.M. 1998. Combining change-in-ratio, index-removal, and removal models for estimating population size. *Biometrics*, **54**, 815–872.

Cooke, J.G. 1997. An implementation of a surfacing-based approach to abundance estimation of minke whales from shipborne surveys. *Report of the International Whaling Commission*, **47**, 513–528.

Cooke, J.G. unpublished. *A modification of the radial distance method for dual-platform line-transect analysis, to improve robustness.* Paper SC/53/IA31 presented to the International Whaling Commission Scientific Committee, 2001.

Cooke, J.G., and Leaper, R. unpublished. *A general modelling framework for the estimation of whale abundance from line transect surveys.* Paper SC/50/RMP21 presented to the International Whaling Commission Scientific Committee, 1998.

Cormack, R.M. 1989. Log-linear models for capture-recapture. *Biometrics*, **45**, 395–413.

Cox, R.D., and Hinkley, D.V. 1974. *Theoretical statistics.* London: Chapman & Hall.

Cressie, N.A.C. 1998. *Statistics for spatial data.* New York: Wiley.

Davison, A.C., and Hinkley, D.V. 1997. *Bootstrap methods and their application.* Cambridge: Cambridge University Press.

Diggle, P.J. 1983. *Statistical analysis of spatial point patterns.* London: Academic Press.

Drummer, T.D., and McDonald, L.L. 1987. Size bias in line transect sampling. *Biometrics*, **43**, 13–21.

Efron, B., and Tibshirani, R.J. 1993. *An introduction to the bootstrap.* New York: Chapman & Hall.

El Khorazaty, M.N., Imrey, P.B., Koch, G.O., and Wells, H.B. 1977. A review of methodological stratiegies for estimating the total number of events with data from multiple-record systems. *International Statistical Review*, **45**, 129–157.

Evans, M.A., Bonett, D.G., and McDonald, L.L. 1994. A general theory for modeling capture-recapture data from a closed population. *Biometrics*, **50**, 396–405.

Feinberg, S.E. 1972. The multiple-recapture census for closed populations and incomplete 2^k contingency tables. *Biometrika*, **59**, 591–603.

Godambe, V.P. 1985. The foundations of finite sample estimation in a stochastic process. *Biometrika*, **72**, 419–428.

Gunderson, D.R. 1993. *Surveys of fisheries resources.* New York: Wiley.

Hastie, T.J., and Tibshirani, R.J. 1990. *Generalized additive models.* London: Chapman & Hall.

Hedley, S.L. 2000. *Modelling heterogeneity in cetacean surveys.* PhD Thesis, University of St Andrews, St Andrews.

Hedley, S.L., Buckland, S.T., and Borchers, D.L. 1999. Spatial modelling from line transect data. *Journal of Cetacean Research and Management,* 1, 255–264.

Hiby, L., and Lovell, P. 1998. Using aircraft in tandem formation to estimate abundance of harbour porpoise. *Biometrics*, **54**, 1280–1289.

Horvitz, D.G., and Thompson, D.J. 1952. A generalization of sampling without replacement from a finite universe. *Journal of the American Statistical Association*, **47**, 663–685.

Huggins, R.M. 1989. On the statistical analysis of capture experiments. *Biometrika*, **76**, 133–140.

Huggins, R.M., and Yip, P.S.F. 1997. Statistical analysis of removal experiments with the use of auxiliary variables. *Statistica Sinica*, **7**, 705–712.

Hunter, J.R., and Lo, N.C.H. 1993. Ichthyoplankton methods for estimating fish biomass: introduction and terminology. *Bulletin of Marine Science,* **53**, 723–727.

Kalman, R.E. 1960. A new approach to linear filtering and prediction problems. *Transactions of ASME Journal of Basic Engineering*, **82**, 35–45.

Kotz, S., and Johnson, N.L. (eds). 1982. *Encyclopedia of statistical sciences.* Vol. 2. New York: Wiley.

Laake, J.L. 1999. Distance sampling with independent observers: Reducing bias from heterogeneity by weakening the conditional independence assumption. *Pages 137–148 of:* Garner, G.W., Amstrup, S.C., Laake, J.L., Manly, B.F.J., McDonald, L.L., and Robertson, D.G. (eds), *Marine mammal survey and assessment methods.* Rotterdam: Balkema.

Lee, S.M., and Chao, A. 1994. Estimating population size via sample coverage for closed capture-recapture models. *Biometrics*, **50**, 88–97.

Lloyd, C.J. 1987. Optimal maringale estimating equations in a stochastic process. *Statistics and Probability Letters*, **5**, 381–387.

Manly, B.F.J. 1997. *Randomization, bootstrap and Monte Carlo methods in biology.* 2nd edn. London: Chapman & Hall.

Manly, B.F.J, McDonald, L.L., and Garner, G.W. 1996. Maximum likelihood estimation for the double-count method with independent observers. *Journal of Agricultural, Biological and Environmental Statistics*, **1**, 170–189.

Marques, F.F.C., and Buckland, S.T. 2003. Incorporating covariates into standard line transect analyses. *Biometrics*, 924–935.

Meyer, R., and Millar, R.B. 1999a. Bayesian stock assessment using a state-space implementation of the delay difference model. *Canadian Journal of Fisheries and Aquatic Sciences*, **56**, 37–52.

Meyer, R., and Millar, R.B. 1999b. BUGS in Bayesian stock assessments. *Canadian Journal of Fisheries and Aquatic Sciences*, **56**, 1078–1087.

Millar, R.B., and Meyer, R. 2000a. Bayesian state-space modeling of age-structured data: fitting a model is just the beginning. *Canadian Journal of Fisheries and Aquatic Sciences*, **57**, 43–50.

Millar, R.B., and Meyer, R. 2000b. Non-linear state space modelling of fisheries biomass dynamics by using Metropolis-Hastings within-Gibbs sampling. *Applied Statistics*, **49**, 327–342.

Newman, K.B. 1998. State-space modeling of animal movement and mortality with application to salmon. *Biometrics*, **54**, 1290–1314.

Newman, K.B. 2000. Heirarchic modeling of salmon harvest and migration. *Journal of Agricultural, Biological and Environmental Statistics*, **5**, 430–455.

Norris, J.L., and Pollock, K.H. 1995. A capture-recapture model with heterogeneity and behavioural response. *Environmental and Ecological Statistics*, **2**, 305–313.

Norris, J.L., and Pollock, K.H. 1996. Nonparametric MLE under two closed capture-recapture models with heterogeneity. *Biometrics*, **52**, 639–649.

Otis, D.L., Burnham, K.P., White, G.C., and Anderson, D.R. 1978. Statistical inference from capture data on closed animal populations. *Wildlife Monographs*, **62**, 1–135.

Palka, D. 1995. Abundance estimate of the Gulf of Maine harbour porpoise. *Report of the International Whaling Commission*, **Special Issue 16**, 27–50.

Pledger, S. 2000. Unified maximum likelihood estimates for closed capture-recapture models using mixtures. *Biometrics*, **56**, 434–442.

Pollock, K.H. 1974. *The assumption of equal catchability of animals in tag-recapture experiments.* PhD Thesis, Cornell University, New York.

Pollock, K.H., and Otto, M.C. 1983. Robust estimation of population size in closed animal populations from capture-recapture experiments. *Biometrics*, **39**, 1035–1050.

Pollock, K.H., Hines, J.E., and Nichols, J.D. 1984. The use of auxiliary variables in capture-recapture and removal experiments. *Biometrics*, **40**, 329–340.

Ramsey, F.L., Wildman, V., and Engbring, J. 1987. Covariate adjustments to effective area in variable-area wildlife surveys. *Biometrics*, **43**, 1–11.

Reilly, S.B., Rice, D.W., and Wolman, A.A. 1980. Preliminary population estimate for the California gray whale based upon Monterey shore censuses. *Report of the International Whaling Commission*, **30**, 359–368.

Reilly, S.B., Rice, D.W., and Wolman, A.A. 1983. Population assessment of the gray whale, *Eschrichtius robustus*, from California shore censuses, 1967–80. *Fishery Bulletin*, **81**, 267–281.

Sanathanan, L. 1972. Estimating the size of a multinomial population. *Annals of Mathematical Statistics*, **43**, 142–152.

Schwarz, C.J., and Seber, G.A.F. 1999. Estimating animal abundance: Review III. *Statistical Science*, **14**, 427–456.

Schweder, T. 1990. Independent observer experiments to estimate the detection function in line transect surveys of whales. *Report of the International Whaling Commission*, **40**, 349–356.

Schweder, T., Skaug, H.J., Langaas, M., and Dimakos, X.K. 1999. Simulated likelihood methods for complex double-platform line transect surveys. *Biometrics*, **55**, 678–687.

Seber, G.A.F. 1970. The effects of trap response on tag-recapture estimates. *Biometrika*, **26**(57), 13–22.

Seber, G.A.F. 1982. *The estimation of animal abundance and related parameters.* 2nd edn. London: Charles Griffin.

Silvey, S.D. 1975. *Statistical inference.* London: Chapman & Hall.

Skaug, H.J., and Schweder, T. 1999. Hazard models for line transect surveys with independent observers. *Biometrics*, **55**, 29–36.

Thompson, S.K. 1992. *Sampling.* New York: Wiley.

Thompson, S.K., and Seber, G.A.F. 1996. *Adaptive sampling.* New York: Wiley.

Trenkel, V.M., Elston, D.A., and Buckland, S.T. 2000. Fitting population dynamics models to count and cull data using sequential importance sampling. *Journal of the Americal Statistical Association*, **95**, 363–374.

Udevitz, M.S., and Pollock, K.H. 1995. Using effort information with change-in-ratio data for population estimation. *Biometrics*, **51**, 471–481.

Wittes, J.T. 1972. On the bias and estimated variance of Chapman's two-sample capture-recapture population estimate. *Biometrics*, **28**, 592–597.

Wood, S.N. 2001. Partially specified ecological models. *Ecological Mongraphs*, **71**, 1–25.

Index